Advances in Intelligent and Soft Computing

152

Editor-in-Chief

Prof. Janusz Kacprzyk
Systems Research Institute
Polish Academy of Sciences
ul. Newelska 6
01-447 Warsaw
Poland
E-mail: kacprzyk@ibspan.waw.pl

T0137887

For further volumes:
http://www.springer.com/series/4240

Pierpaolo Vittorini, Rosella Gennari,
Ivana Marenzi, Fernando De la Prieta,
and Juan M. Corchado Rodríguez (Eds.)

International Workshop on Evidence-Based Technology Enhanced Learning

 Springer

Editors
Pierpaolo Vittorini
Department of Internal Medicine
and Public Health
University of L'Aquila
Italy

Rosella Gennari
Computer Science Faculty
Free University of Bozen-Bolzano
Bolzano
Italy

Ivana Marenzi
L3S Research Center
Leibniz University of Hanover
Hannover
Germany

Fernando De la Prieta
Department of Computing Science
University of Salamanca
Salamanca
Spain

Juan M. Corchado Rodríguez
Department of Computing Science
University of Salamanca
Salamanca
Spain

ISSN 1867-5662
ISBN 978-3-642-28800-5
DOI 10.1007/978-3-642-28801-2
Springer Heidelberg New York Dordrecht London

e-ISSN 1867-5670
e-ISBN 978-3-642-28801-2

Library of Congress Control Number: 2012933102

© Springer-Verlag Berlin Heidelberg 2012
This work is subject to copyright. All rights are reserved by the Publisher, whether the whole or part of the material is concerned, specifically the rights of translation, reprinting, reuse of illustrations, recitation, broadcasting, reproduction on microfilms or in any other physical way, and transmission or information storage and retrieval, electronic adaptation, computer software, or by similar or dissimilar methodology now known or hereafter developed. Exempted from this legal reservation are brief excerpts in connection with reviews or scholarly analysis or material supplied specifically for the purpose of being entered and executed on a computer system, for exclusive use by the purchaser of the work. Duplication of this publication or parts thereof is permitted only under the provisions of the Copyright Law of the Publisher's location, in its current version, and permission for use must always be obtained from Springer. Permissions for use may be obtained through RightsLink at the Copyright Clearance Center. Violations are liable to prosecution under the respective Copyright Law.
The use of general descriptive names, registered names, trademarks, service marks, etc. in this publication does not imply, even in the absence of a specific statement, that such names are exempt from the relevant protective laws and regulations and therefore free for general use.
While the advice and information in this book are believed to be true and accurate at the date of publication, neither the authors nor the editors nor the publisher can accept any legal responsibility for any errors or omissions that may be made. The publisher makes no warranty, express or implied, with respect to the material contained herein.

Printed on acid-free paper

Springer is part of Springer Science+Business Media (www.springer.com)

Preface

Research on Technology Enhanced Learning (TEL) investigates how information and communication technologies can be designed in order to support pedagogical activities. The Evidence Based Design (EBD) of a system bases its decisions on empirical evidence and effectiveness. The evidence-based TEL workshop (ebTEL) brings together TEL and EBD. The workshop proceedings collects contributions concerning evidence based TEL systems, like their design following EBD principles as well as studies or best practices that educators, education stakeholders or psychologists used to diagnose or improve their students' learning skills, including students with specific difficulties (e.g. poor/slow readers, students living in impoverished communities or families).

The ebTEL'12 workshop was launched under the collaborative frame provided by the European TERENCE project (http://www.terenceproject.eu). The TERENCE project, n. 257410, is funded by the European Commission through the Seventh Framework Programme for Research and Technological Development, Strategic Objective ICT-2009.4.2: ICT: Technology-enhanced learning. TERENCE is building an AI-based Adaptive Learning System (ALS) for reasoning about stories, in Italian and in English, through reading comprehension interventions in the form of smart games. The project also aims at developing innovative usability and evaluation guidelines for its users. The guidelines and the ALS will result from a cross-disciplinary effort of European experts in diverse and complementary fields (art and design, computers, engineering, linguistics and medicine), and with the constant involvement of end-users (persons with impaired hearing and their educators) from schools in Brighton (Great Britain), and from Veneto region (Italy).

The ebTEL workshop invited authors in the area of TEL, both from computer science as well as evidence-based medicine, educational psychology and pedagogy. In this manner, the international ebTEL'12 workshop wants to be a forum in which TEL researchers and practitioners alike can discuss ideas, projects, and lessons related to ebTEL. The workshop takes place in Salamanca, Spain, on March 28th-30th 2012.

This volume presents the papers that were accepted for the first edition of ebTEL. The full program contains 16 selected papers (3 in the Psychology and Pedagogy Track and 13 in the Information and Communication Technology Track) from 6 countries (Germany, Italy, Romania, Saudi Arabia, Spain, United Kingdom, USA). Each paper was reviewed by, at least, two different reviewers, from an international committee composed of 27 members of 5 countries. The quality of papers was on average good, with an acceptance rate of 76.2%.

We would like to thank all the contributing authors, the reviewers, the sponsors (IEEE Systems Man and Cybernetics Society Spain, AEPIA Asociación Española para la Inteligencia Artificial, APPIA Associação Portuguesa Para a Inteligência Artificial, CNRS Centre national de la recherché scientifique and STELLAR), as well as the members of the Program Committees, TERENCE members and the Organising Committee for their hard and highly valuable work. Special thanks are due to Wolfgang Nejdl of the L3S Research Center of the University of Hannover (Germany), invited speaker of the workshop with the talk "Web Science @ L3S – Interdisciplinary Research Challenges". The work of all these people contributed to the success of the first edition of the ebTEL'12workshop. Thank you all for your help, ebTEL'12 would not exist without your contribution.

The Editors
Pierpaolo Vittorini
Rosella Gennari
Ivana Marenzi
Fernando De la Prieta
Juan M. Corchado Rodríguez

Organization

Steering Committee

Juan M. Corchado	University of Salamanca, Spain
Fernando De la Prieta	University of Salamanca, Spain
Rosella Gennari	Free University of Bozen-Bolzano, Italy
Ivana Marenzi	L3S Research Center, Germany
Pierpaolo Vittorini	University of L'Aquila, Italy

Program Committee

Fabian Abel	Delft University of Technology, Nederland
Tomaz Amon	AMNIN, Slovenia
Anthony Baldry	University of Messina, Italy
Juan M. Corchado	University of Salamanca, Spain
Fernando De la Prieta	University of Salamanca, Spain
Juan F. De Paz Santana	University of Salamanca, Spain
Dina Di Giacomo	University of l'Aquila, Italy
Tania Di Mascio	University of l'Aquila, Italy
Rosella Gennari	Free University of Bozen-Bolzano, Italy
Ana Belén Gil	University of Salamanca, Spain
Óscar Gil	University of Salamanca, Spain
Nicola Henze	L3S, Leibniz University of Hanover, Germany
Eelco Herder	L3S, Leibniz University of Hanover, Germany
Ralf Kerstel	L3S, Leibniz University of Hanover, Germany
Ralf Klamma	RWTH Aachen University, Germany
Rita Kupetz	Leibniz University Hanover, Germany
Fabrizio Maggi	University of Pavia, Italy
Ivana Marenzi	L3S, Leibniz University of Hanover, Germany
Maria Moreno Jaén	University of Granada, Spain
Stefano Necozione	University of l'Aquila, Italy
Emanuele Pianta	FBK-irst, Italy
Sara Rodríguez	University of Salamanca
Maria Grazia Sindoni	University of Messina, Italy
Marcus Specht	Open University of the Netherlands
Sara Tonelli	FBK-irst, Italy
Pierpaolo Vittorini	University of l'Aquila, Italy
Carolina Zato	University of Salamanca, Spain

Local Organising Committee

Juan M. Corchado (Chairman)	University of Salamanca, Spain
Javier Bajo	Pontifical University of Salamanca, Spain
Sara Rodríguez	University of Salamanca, Spain
Dante I. Tapia	University of Salamanca, Spain
Emilio Corchado	University of Salamanca, Spain
Juan F. De Paz	University of Salamanca, Spain
Fernando De la Prieta	University of Salamanca, Spain
Davinia Carolina Zato Domínguez	University of Salamanca, Spain
Óscar Gil Gonzalo	University of Salamanca, Spain

Contents

Understanding Computer Architecture with Visual Simulations: What Educational Value?

Besim Mustafa and Peter Alston

Abstract. Many software simulators have been created for educational purposes. Such educational tools need to be both engaging and pedagogically sound if they are to enhance students' learning experiences. A system simulator for computer architecture teaching and learning has been developed and refined according to both established and emerging new principles of pedagogy. A methodology for evaluating the simulator's educational value has been identified and successfully applied during scheduled practical tutorial classes in years two and three of a three year undergraduate computing degree. The results are presented both qualitatively and quantitatively and are strongly indicative of the positive pedagogical value offered by the system simulations.

1 Introduction

We have been designing and delivering undergraduate teaching modules on computer architectures and operating systems for the past seven years. In order to support the practical tutorial sessions we developed an integrated set of simulators collectively identified as the system simulator [5]. The simulators conform to modern principles of pedagogy with respect to facilitating student engagement [6] and enhancing student learning experiences [1]. This paper presents examples of practical assignments using the simulations in order to demonstrate the ways in which the above principles are adhered to in the design of the simulations. We also present a brief account of a quantitative method of evaluating the educational value of the simulations and the ways in which the measurements were defined, gathered and analysed.

The integrated simulators demonstrate the essence of modern computer architecture, by capturing the following key aspects of computer technology in one educational software package [9]: the generation of CPU instructions by

Besim Mustafa · Peter Alston
Business School, Faculty of Arts and Sciences, Edge Hill University, Ormskirk, UK
e-mail: {mustafab,alstonp}@edgehill.ac.uk

P. Vittorini et al. (Eds.): International Workshop on Evidence-Based TEL, AISC 152, pp. 1–9.
springerlink.com   © Springer-Verlag Berlin Heidelberg 2012

assemblers and compilers; the CPU as the processor of the instructions; the operating system (OS) as the facilitator of multiprocessing and multi-threading of the CPU instructions.

The CPU simulator simulates a fictitious, but highly realistic, RISC type CPU architecture. It has a small set of CPU instructions and a register file with selectable number of registers. This simulator incorporates a five-stage pipeline simulator as well as data and instruction cache simulators. The pipeline simulator supports operand forwarding and jump prediction using a jump table. The cache simulators support cache organizations, placement and replacement policies as well as cache coherency policies. Cache simulators display graphical performance statistics.

The OS simulator supports two main aspects of a computer system's resource management: process management and memory management. The process scheduler supports priority-based, pre-emptive and round-robin scheduling. Threads are supported via teaching language constructs enabling students to explore mutual exclusion, process synchronization and deadlock concepts. The virtual memory simulations explore address translation, placement and replacement concepts.

2 Related Work

Modern computer architectures are examples of co-operating systems across tightly coupled interfaces. For example, there are design considerations in hardware in order to assist the operating system and vice versa and highly-optimized compilers include technology in order to assist the hardware and vice versa. It follows that a simulator for computer architecture education should also reflect this interplay and interdependency between computer hardware and software. We have not come across such an educational simulator and therefore we were motivated to develop our own.

Educational simulators often originate from different sources, look and feel different and simulate isolated cases of computer technology in ad hoc and piece-meal fashion using artificial data [3,7,8,11,12]. This makes them unsuitable for studying how different parts of a system fit and work together in educational settings. Moreover, apart from the efforts to evaluate isolated algorithm simulations [7,10], we did not find any references to methodical evaluation of more complex system simulations for their pedagogical value. This paper offers one such methodology.

3 The Teaching and Learning Strategy

We have successfully integrated the system simulator into our modules on computer architecture and operating systems. During each practical tutorial session the students work in small groups. The practical exercises are available online and are designed to encourage critical thinking and deeper understanding of the theory. Year two students study advanced computer architecture and concentrate on performance issues. They study caches, pipelines and compiler optimizations. Year three students study internals of operating systems and explore scheduling mechanisms, memory management techniques, threads, deadlocks and synchronization.

Each week a one-hour lecture is followed by a two-hour tutorial session. The simulation exercises are completed during each tutorial session and over a semester the simulations cover a broad range of computer architecture topics. At the end of the semester the completed portfolio of exercises are assessed.

4 Sample Simulation Tutorial Examples

In this section we look at examples of practical tutorials taken from *Advanced Computer Architecture* and *Operating Systems* modules in which the students use the integrated simulators as part of their coursework portfolios. All examples below rely on the CPU simulator executing instructions, generated by the inbuilt compiler or manually entered, in the background as in real computer systems.

During the tutorial on cache technology the students investigate different cache types, cache sizes, block and set sizes as well as the placement and replacement policies. They study the advantages and disadvantages of directly-mapped and 2-way/4-way/8-way set-associative cache organizations. They use the inbuilt compiler to create programs that write and read arrays of bytes and run the programs on different sizes of cache on the above cache organizations and record the cache performances. They then comment on their observations. Finally they devise experiments to show the impact of programming style on cache performance. Fig 1 shows the data cache simulator.

The tutorial on CPU pipelines allows the students to switch the CPU pipeline on or off and pipeline optimizations such as operand forwarding and jump prediction can be selected or deselected. The simulator displays colour-coded stages of instructions as they progress through the pipeline, calculates and displays the average clocks per instruction (CPI) and the speed-up factor (SF). The students run programs on various configurations of the pipeline (e.g. hazard bubbles on and off, use NOP instructions to delays, optimizations on and off) and record the CPI and the SF. They study the impact of loop-unrolling and out-of-order instruction optimizations on pipeline performance.

The tutorial on threads uses the inbuilt compiler where the threads are implemented as system calls via a language construct that identifies code in selected subroutines as part of a thread. Calls to these subroutines invoke the OS simulator to create the threads. Students explore the concepts of critical regions, synchronization and counting semaphores for protecting shared global resources. They explore parent/children hierarchies and demonstrate client/server software using multiple threads when processing client requests in separate threads.

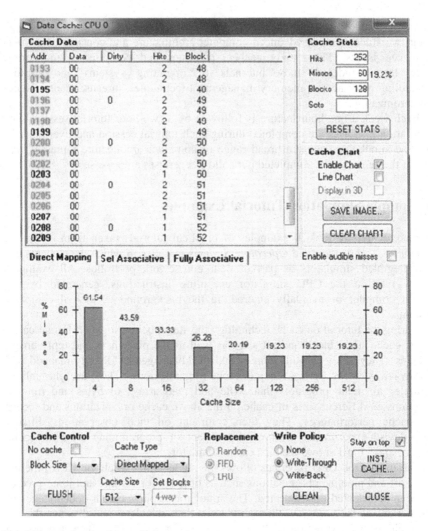

Fig. 1 Data cache simulator window

5 Evaluation Methodology

We used a two-pronged strategy in evaluating the simulations. We first evaluated the simulations by mapping the scope of the simulations to the *Engagement Taxonomy* [6] and to Bloom's *Learning Taxonomy* [1]. The former is developed as a measure of the effectiveness of algorithm visualizations, whilst the latter is based on learning theory and provides a measure of the depth of learning. We next used the quasi-experimental method to gather and analyze the quantitative data.

5.1 Taxonomies of Pedagogy and Qualitative Evaluation

In this section we show the degree of conformance of the simulations to the two taxonomies by mapping various features of the simulations to the individual elements of the taxonomies as a measure of the effectiveness of the simulations. It is interesting to note that there is some common ground between the two mappings shown in Tables 1 and 2. It can be seen that the simulations are able to fully support the requirements of the two taxonomies across all their elements qualitatively indicating that they have the capacity to both gainfully engage and educate.

Table 1 Mapping onto the Engagement Taxonomy

Attributes of Engagement Taxonomy	View	Can view data held in cache, its address, no of hits against each entry, block numbers, set numbers, etc. Colour coded bar charts display in real-time data misses and hits.
	Respond	Responding is mainly guided by the instructions in the exercise sheets; students respond to questions and make notes on their observations; simulations don't require specific responses.
	Change	Can change pipeline status: pipeline enabled or disabled; pipeline optimizations selected or deselected; pipeline history recording switched on or off; pipeline history played back or forward and can be reset. Jump table contents can be cleared.
	Construct	Students use the inbuilt compiler and the language to construct scenarios where the style of programming can affect the efficient use of the cache by writing assembly code to highlight and demonstrate strengths and weaknesses of cache mappings.
	Present	Students can present the implementation of "consumers" and "producers" problem and solution using threads and semaphores. Students construct a simple multi-threaded server program and use it to demonstrate basics of client request handling.

Table 2 Mapping onto Bloom's Taxonomy

Bloom's Levels of Learning Taxonomy	Knowledge	Knowing of different cache organizations. Awareness of hit rate as a measure of performance. Be able to explain meanings of sets and blocks.
	Comprehension	Understanding the use of NOP instructions to compensate for data hazards. Appreciation of the role of compiler optimizations on pipeline performance. Ability to calculate CPI and SF and understand how they are related.
	Application	Application of compiler optimization loop-unrolling to increase pipeline performance. Designing test cases to demonstrate different jump instruction's impact on pipeline jump prediction mechanism.
	Analysis	Experimenting with set associative mapping using 2-way, 4-way and 8-way sets and comparing relative merits. Students can experiment with write-through and write-back policies and comment on which is best and under what conditions.
	Synthesis	Using inbuilt compiler and assembler to create CPU instructions to demonstrate pipeline features and show how software can assist hardware in increasing system performance.
	Evaluation	Evaluating the relative merits of set associative cache organization's 2-way to 8-way sets and evaluate the optimum value. Running a program and evaluating graphs of miss rates against cache sizes for all cache organizations.

5.2 The Quantitative Evaluation

The primary data gathering was conducted in the normal course of the scheduled tutorial sessions for which no special arrangements were made. The only additions to the tutorial sessions were the short pre and post test questions (each lasting no more than 10 minutes to do) before and after the simulation exercises.

The quantitative evaluation required effective and meaningful measurements with careful attention to both the nature of the measurements and the processes in which they were taken if this was to yield reliable results. For this we first identified and then applied a set of guidelines to assist us. Table 3 presents these guidelines. We then applied them to defining the learning outcomes and designing the pre and post tests.

Table 3 Guidelines for collecting evaluation data

Making sure what is measured is what is being learned	Define clear and achievable learning outcomes (LOs); target for a small number of learning outcomes, say 4; design practical exercises closely aligned with the learning outcomes.
Making sure a meaningful baseline is established prior to measuring any changes	Use pre test questions that reflect the general goals of learning outcomes; seek evidence for levels of confidence; ask one question directly corresponding to each learning outcome.
Making sure measurement in any change is done only against the established baseline	Use post test questions that correspond to pre test questions in modified form, one question against each learning outcome; seek evidence of confidence attributable to the intervention.
Making sure what is measured is only due to the intervention	Do post tests immediately after the intervention; link questions to the method of intervention; seek level of confidence in the use of the intervention alone.

The pre and post tests took the form of 5-point Likert scale (strongly agree, agree, not sure, disagree, strongly disagree) with five items in each test (Q1 to Q5). The items of the tests took the form of confidence-based opinions rather than multiple choice questions. Table 4 shows examples of LOs and the corresponding styles of pre and post test questions we used that conform to the guidelines above.

The pre tests were administered soon after the lecture and prior to the practical exercises. The post tests were carried out straight after the completion of the practical exercises. This method afforded increased confidence in safely attributing the differences in the pre and post results to the learning intervention.

6 Results and Analysis

Due to relatively small sample sizes and because the data is not normally distributed we opted for the non-parametric statistical analysis Wilcoxon signed-rank test [2] used to compare two dependent conditions originating from the same set of students. Table 5 shows the analysis results of the data. We regard values of $p < 0.05$ as statistically significant. These results strongly suggest that the simulations have indeed been responsible for increased student confidence in the understanding of the theoretical concepts and their practical applications.

Table 4 Examples of pre and post test questions

	Cache Simulator Pre/Post Tests	Pipeline Simulator Pre/Post Tests
LO	Investigate 2-way and 4-way set-associative cache mappings	Describe a pipeline technique to eliminate data hazard
Pre	I understand and can explain how set associative cache organizations function	I can describe a method used by CPU pipeline in order to eliminate data hazards
Post	The simulations helped me understand and explain how set-associative cache organizations function	The tutorial exercises helped me understand and describe a method used by CPU pipeline to eliminate data hazards
LO	Explain the effect of cache size and mapping scheme on cache performance	Demonstrate compiler loop unrolling and out-of-order optimizations' benefits for instruction pipelining
Pre	I understand and can explain the effects of the cache type and the cache size on cache performance	I understand and can describe two methods modern compilers use in order to minimize CPU pipeline hazards
Post	The simulator aided me in demonstrating and understanding the effects of cache type and cache size on cache performance	The simulations aided my understanding of two methods compilers employ to minimize CPU pipeline hazards

We then conducted a reliability analysis on our pre and post tests which yielded the Cronbach's alpha as the measure of test reliability [2]. Table 5 shows these values for both the pre and the post tests. The values are indicative of high degree of reliability and internal consistency in our tests.

Table 5 Statistical analysis of evaluation data

Tutorial Exercise	Sample size	Value of p					Reliability,	
		Q1	Q2	Q3	Q4	Q5	Pre Test	Post Test
Investigating CPU Cache	12	0.015	0.011	0.003	0.003	0.002	0.898	0.770
Investigating CPU Pipeline	12	0.180	0.002	0.020	0.008	0.008	0.849	0.883
Investigating Threads	18	0.002	0.001	0.000	0.002	0.001	0.935	0.897
Investigating Synchronization	24	0.000	0.000	0.000	0.000	0.000	0.875	0.897

7 Conclusions

Blooms taxonomy of learning has been with us for a number of years and is considered to be a 'tried and tested' method when gauging what level learners are at in their learning journey. The Engagement taxonomy focuses on how these learners are 'involved' and interact with the learning environment through the use of visualizations and simulations requiring and encouraging active learner engagement; it provides unified guidelines that can help shape the visual learning tools.

One of the difficult tasks that learners face is the ability to become immersed in a subject that they may not fully understand which is heavily based on theory and is difficult to visualize. The likely answer here is that content is being delivered which is not appropriate to the learner's individual learning style [4]. This could in turn have an effect on their ability to fully engage with assessment tasks in order to demonstrate their competence. All learners are individuals in their ability to learn and as educators we should be striving to help learners achieve this 'individualized learning journey'.

The methods and results that are presented in this paper indicate that the use of the system simulator we have developed does help the learners in understanding the material they need in order to meet the learning outcomes of their modules. What is more interesting is not only the similarity between the two taxonomies which shows us that students are learning and engaging with the learning material, but we are also catering for the visual and kinesthetic learners and almost providing an individualized learning journey for them. Our next focus is to see how we can also cater for the auditory learners and for those with learning difficulties.

References

[1] Bloom, B.S., Krathwohl, D.R.: Taxonomy of Educational Objectives; the Classification of Educational Goals, Handbook I: Cognitive Domain. Addison-Wesley (1956)
[2] Field, A.: Discovering Statistics Using SPSS, 3rd edn. SAGE Publications Ltd (2009)
[3] Garrido, J.M., Schlesinger, R.: Principles of Modern Operating Systems. Jones and Bartlett (2008)
[4] Leite, W.L., Svinicki, M., Shi, Y.: Attempted Validation of the Scores of the VARK: Learning Styles Inventory With Multitrait–Multimethod Confirmatory Factor Analysis Models, p. 2. SAGE Publications (2009)
[5] Mustafa, B.: YASS: A System Simulator for Operating System and Computer Architecture Teaching and Learning. In: FISER 2009 Conference, Famagusta, Cyprus, March 22-24 (2009)
[6] Naps, T.L., Fleischer, R., McNally, M., et al.: Exploring the Role of Visualization and Engagement in Computer Science Education. ACM SIGCSE Bulletin 35(2) (June 2003)
[7] Nikolic, B., Radivojevic, Z., Djordjevic, J., Milutinovic, V.: A Survey and Evaluation of Simulators Suitable for Teaching Courses in Computer Architecture and Organization. IEEE Transactions on Education 52(4) (November 2009)
[8] Null, L., Lobur, J.: MarieSim: The MARIE computer simulator. ACM Journal of Educational Resources in Computing 3(2), article 1 (June 2003)
[9] Teach-sim educational simulators, http://teach-sim.com (accessed November 24, 2011)

[10] Urquiza-Fuentes, J., Valezquez-Iturbide, J.A.: A survey of Successful Evaluations of Program Visualization and Algorithm Animation Systems. ACM Transactions on Computing Education 9(2), article 9 (June 2009)

[11] Yehezkel, C., Yurcik, W., Pearson, M., Armstrong, M.: Three simulator tools for teaching computer architecture: EasyCPU, Little Man Computer, and RTLSim. ACM Journal of Educational Resources in Computing 1(4), 60–80 (2002)

[12] Yurcik, W., Wolffe, G.S., Holliday, M.A.: A survey of simulators used in computer organization/architecture courses. In: SCSC 2001, Orlando FL, USA (July 2001)

Evaluation Plan of TERENCE: When the User-Centred Design Meets the Evidence-Based Approach

Vincenza Cofini, Dina Di Giacomo, Tania Di Mascio,
Stefano Necozione, and Pierpaolo Vittorini

Abstract. TERENCE is an FP7 EU project that aims at developing an adaptive learning system with the twofold objective of helping children in improve deep text understanding, and supporting teachers in their daily work. The present paper focuses on the design of the evaluation of the pedagogical effectiveness and the usability of the TERENCE software. It starts from the user-centred design experience, evidence-based medicine, psychology, and from discussions about statistical methods and ethics considerations. The objective is to provide an innovative, evidence-based and efficient support, for children and teachers, that could be an efficient alternative to the traditional method of reading, so as to prevent and reduce problems of text comprehension that represent a public health and social problem. For this purpose, we developed an evaluation protocol within a reading laboratory in collaboration with teachers, to be hosted in the school structures that will join the project in Italy.

Keywords: EBM, UCD, TEL.

1 Introduction

From the age of 7-8 until the age of 11, children develop as independent readers. Nowadays, more and more children in that age range turn out to be *poor comprehenders*, i.e. they have well developed low-level cognitive skills (e.g. vocabulary

Vincenza Cofini · Dina Di Giacomo · Stefano Necozione ·
Pierpaolo Vittorini
University of L'Aquila, Dep. of Internal Medicine and Public Health, Via S. Salvatore, 67100 L'Aquila, Italy

Tania Di Mascio
University of L'Aquila, Department of Electrical and Information Engineering
Via G Gronchi 18, 67100 L'Aquila, Italy

P. Vittorini et al. (Eds.): International Workshop on Evidence-Based TEL, AISC 152, pp. 11–18.
springerlink.com
© Springer-Verlag Berlin Heidelberg 2012

knowledge [2, 8]), though they have difficulties in deep text comprehension (e.g. inference making [13]). TERENCE (An AdapTivE LeaRning SystEm for ReasoNing about Stories with Poor Comprehenders and their Educators) is an FP7 ICT project that aims at designing, developing and *evaluating* the first Adaptive Learning System (ALS) for poor comprehenders. In particular, it is an innovative story-telling web tool whose dual purpose is to improve the comprehension ability of written text in poor comprehenders and to support their teachers. Evaluating TERENCE require considering both the pedagogical effectiveness of the overall system and its software usability, articulated through a "small scale pedagogical evaluation", a "small scale evaluation adequacy-satisfaction" and a "large scale evaluation". These research evaluation of activities, are designed and developed with the direct involvement of TERENCE's users at school. For this purpose, we developed a protocol for a *reading laboratory* in collaboration with teachers, to be hosted in the school structures that will join the project in Italy. The paper represents the result of a collaboration among multidisciplinary competences. We started from User Centered Design (UCD) experience, Evidence-Based Medicine (EBM), psychology, and from discussion about statistical methods and ethics considerations. In particular, the paper presents the results of the integration of UCD and EBM techniques, in the optics of providing solid basis for evaluating the TERENCE software from the twofold perspective of a pedagogical effectiveness and usability. The EBM is the conscientious, explicit, and judicious use of current best evidence in making decisions about the care of individual patients. The practice of evidence based medicine means integrating individual clinical expertise with the best available external clinical evidence from systematic research [10]. Otherwise, the UCD is an Human Computer Interaction (HCI) methodology placing the end user, user organisations and support teams at the centre of the design and evaluation processes. This means that the system's users are involved from the very beginning of the project, and can participate in the design and evaluation of the system. The UCD methodology can thus be defined as "a process focusing on usability throughout the entire development process and further throughout the system life cycle" [6]. First of all, the paper lists the organizational constraints emerged from the results of the analysis of the context of use (Section 2), then describes the experiment design with its objectives and how the organizational constraints affected both the proposed methodology and the evaluation goals (Section 3). Finally, the paper ends with a discussion of the lesson learned in using a "mixed" UCD/EBM approach (Section 4).

2 Organisational Constraints

As known, the organizational constraints of the context of use are one of the first activity carried out in UCD during the entire process. It is thus not surprising that this collection is contained in [11] and mainly derives from the advices and suggestions given by teachers both in the first and in the second field studies. Hereinafter, we summarises the organizational constraints derived from the UCD analysis, that guided the design of the EBM approach for the evaluation of TERENCE:

oc1. We observed that parents usually help children in their homework activities. This aspect could represent a bias in the evaluation. *Thus, the evaluation should be conducted in the schools.*

oc2. Even if TERENCE focuses in improving the reading comprehension of poor comprehenders, teachers suggested to include all children of a class, in order not to introduce discriminations. *Thus, the evaluation methodology should include all children.*

oc3. During the field studies, both schools' principals and experts asked us to stay in one classroom for no more than 45 minutes/1 hour, so to preserve the normal lesson's flow, and ensure a proper level of attention (that decreases after that period of time). *Thus, the intervention cannot last more than 45 minutes/1 hour.*

oc4. In general, schools' principals suggested to adequately weight the number of interventions, so to preserve the regularity of the standard school program. *Thus, the intervention should be though as an external activity (as an extra-school lab).*

oc5. We found that public schools, especially in Italy, do not have (i) resources to support further activities than the standard school program; (ii) extra physical environments; (iii) proper devices to run the TERENCE software. *Thus, the management body of TERENCE should provide a support of its own.*

oc6. Clinical experts report that it is very important to take into account the physiological and attentional intervals. *Thus, we should alternate direct stimulations and relaxing activities.*

3   Experiment Design

This section describes the experiment by focusing on the evaluation of the effectiveness and usability of TERENCE in Italy. To this aim, we initially present the objectives of the evaluation (Subsection 3.1). Given the above, a proper methodology is proposed, that tackles the objectives, fits the organizational constraints and is coherent with clinical practice (Subsection 3.2).

3.1   Objectives

The objective of our work is the evaluation of the effectiveness and usability of the TERENCE tool. In order to evaluate the TERENCE system as a whole, we list the specific objectives that contribute to the overall objective concerning the evaluation process: (a) Evaluation of the prior estimated effectiveness with a small scale pedagogical evaluation pre/post TERENCE use; (b) Evaluation of the usability of TERENCE in relation to the interaction between the user and the tool itself, with the small scale evaluation adequacy-satisfaction pre/post TERENCE use; (c) Evaluation of the effectiveness and usability of the TERENCE tool with the large scale evaluation. These specific objectives are reported in subsection 3.3, also with a clear explanation of metrics to measure them.

3.2 Methodology

The organizational constraints described in the previous section guided the following methodological choices, divided into *execution methodology* and *data collection methodology*.

Execution Methodology. The target population is of "all children aged 7-11 years", as discussed in *oc2* of Section 2. Every child can participate, if his/her parents give their informed consent, as the Italian law imposes. Considering the organizational constraint *oc4*, we decided to organise a reading laboratory supporting both traditional methods (paper and pencil method) and the innovative method (interaction with TERENCE), in collaboration with teachers. The laboratory has to be hosted in the school structures that will join the project in Italy, as in *oc1*. Schools principals will inform children, parents and teachers about the project (objective, duration and activities), with the support of the management body of TERENCE, as in *oc5*. The reading laboratory activity is about 45 minutes each time. The session lasts 45 minutes, to guarantee the developing of the activities with high level of attention and concentration, with the respect of the other school activities as well described in *oc3*. During the session, at school, personal trained from the management body of TERENCE is always present, to prepare the activities, to inform the teachers and children about the different steps of the laboratory, to collect data necessary for evaluation, to prepare exercises with the traditional and innovative methods, as in *oc5*. Every child reads stories and is stimulated trough games. In order to take into account *oc6*, games are divided into two categories: (i) smart games to stimulate understanding of the text, and (ii) relaxing games to encourage the consolidation of the learned information. The two type of games are alternated, so to take into account the physiological and attentional intervals, so to improve the interaction with the tool. During the stimulation measurements are collected both by the software and by the management body of TERENCE.

Data Collection Methodology. The data will be collected anonymously, in full compliance with Italian Privacy Laws. Each child will be accompanied by a code that will replace any information that can be traced back to the identity of the child. The management body of TERENCE and teachers will not know the association between code and child. For data collection, and for the preparation of the reading exercises, we will take advantage of a trained person of management body of TERENCE.

3.3 Specific Objectives, Metrics and Methods

This subsection delves into the three specific evaluation objectives mentioned in Subsection 3.1, by initially listing the metrics used to verify their achievement, conceptually divided into psychological, performance and usability metrics:

Psychological

(psy1) *Prove di Lettura MT-2*. The Italian test consents to evaluate the learning ability in the reading and writing in scholar age [4];

(psy2) *P.P.V.T.-R.*. The test enables to evaluate the children ability to understand the lexicon. The test is composed of 180 tables. The subject is asked to point the figure corresponding to the word pronounced by the examiner [5];

(psy3) *Prova di Comunicazione Referenziale*. Aim of this test is the evaluation of the children ability to make and understand simple or complex information. The subject task is to indicate on table or verbally identify the correct information [3];

(psy4) *Coloured Progressive Matrices*. The test consents to evaluate the non-verbal intelligence in developmental age. The test is composed of 36 items and the subject task is to find the correct response between 6 choices on the basis of the target [9];

(psy5) *Batteria per la valutazione della Scrittura e della Competenza Ortografica*. The Italian test evaluates the orthographic ability in developmental age [12];

(psy6) *Neuropsychological Evaluation Battery*. The battery consents to evaluate the neuropsychological ability in the range age 5-11 years old. The cognitive areas investigated are language, memory, attention and logical reasoning [1].

Performances

(per1) *Average word reading time*. The reading time is a parameter computed by the software. It is measured in ms and consists in the time lasted from the visualisation of the TERENCE first page to the last page, divided by the number of words;

(per2) *Average answer time*. The average answer time is a parameter computed by the software. It is measured in ms and consists in the average time to answer to all games of a given story. The time for each game is measured from the visualisation of the game to when the answer is given (decision time + motor time);

(per3) *Accuracy*. The accuracy is a parameter computed by the software. It is a pure number in the range [0,1] and is defined as the ratio between the number of games correctly resolved, and the total number of games. The value is related to one story;

(per4) *Omission ratio*. The omission ratio is a parameter computed by the software. It is a pure number in the range [0,1] and is defined as the ratio between the number of games that were not answered, and the total number of games. The value is related to one story.

Usability

(us1) *Success Rate Per User*. The success rate per user is a parameter computed by the software. It is the percentage of primary tasks successfully completed. It is defined as follows: $S_r = \frac{S+(w \cdot P)}{T}$, where S is the number of primary tasks successfully completed, P is the number of primary tasks partially completed, T is the total number of tasks, and w is a constant in range $[0,1]$, that weights the importance of the tasks partially completed (usually, it is 0.5). For TERENCE, the primary tasks are the following. For the children: avatar selection and setting, book selection, story selection, reading, game session. For the educator: setting of the cognitive variables for the learner, visualisation of the children' progresses, visualisation of the stories read by the learner. For the expert: visualisation of the children' progresses, upload of new stories.

(us2) *Number of useless steps.* Ps_{tot} is the sum of the useless steps per task Ps_t. The number of useless steps per task is the number of steps that are not necessary to complete one of the primary tasks. It is a parameter that is computed by the software. It is a pure number defined as follows: $Ps_{tot} = \sum Ps_t$ where $Ps_t = P_{tot_t} - P_{n_t}$ and P_{tot_t} is the total number of steps for the execution of task t, and P_{n_t} is the number of steps necessary for the execution of task t.

(us3) *User satisfaction.* To measure the satisfaction of the TERENCE users, we will use a self-reported questionnaire with multiple answers, from a Likert scale, whose values are: 0 = Very Dissatisfied, 1 = Dissatisfied, 2 = Average Satisfied, 3 = Satisfied, and 4 = Very Satisfied.

Given the above, the description of the methodological choices for our three main objectives follows.

(a) Evaluation of the prior effectiveness with a small scale pedagogical evaluation pre/post TERENCE use. For this specific objective, the research design is a repeated measures design. Pre/post evaluation designs are best used when measuring outcomes (e.g., knowledge, attitudes, skills, aspirations, behaviours) of programs or activities that are specifically developed to bring them about and you have access to the participants before and after the program. When measuring behaviours using this evaluation design, an appropriate amount of time must be given before the post measurement is administered to allow the behaviour to take place. According to clinical practice, two months seem to be a good time to a complete cycle of the TERENCE use, to study the effect of the method of reading. A random sample is not planned for the small scale evaluation. We aim at working with all children that agree, and a sample of 30 children should suffice to investigate the primary efficacy of TERENCE system. The primary goal is the evaluation of the level of the text understanding after the TERENCE use. The expected result is the improvement of the comprehension text after use of the TERENCE software for every variable studied. We expect: (i) improved scores (psy1-psy6); (ii) lower times (per1-per2); (iii) higher accuracy (per3); (iv) lower omission ratio (per4). The results in the pretest are compared with the results in the post-test with statistical methods (parametric on not), for comparing averages of matched data (pre/post). A p-value of 0.05 will be considered significant.

(b) Evaluation of the usability of TERENCE in relation to the interaction between the user and the tool itself with the small scale evaluation adequacy-satisfaction pre/post TERENCE use. For this specific objective the research design is a repeated measures design. A random sample is not planned for small scale evaluation. We aim at working with all children that agree, and a sample of 30 units should suffice to investigate the primary efficacy of TERENCE system. Children will use the TERENCE method during the laboratory for 2 months, as for objective (a). During the stimulation with TERENCE, we measure the efficiency with (us1-us3). The expected results are the improvement of the interaction with the new method after the laboratory for every studied variable, i.e. an increase of (us1) and a decrease of (us2). For (us3), the user satisfaction will be measured only at the end of the experiment, for descriptive statistics only. Success rate (us1) at time t_0 will be compared with success rate at time t_1, with statistical test for frequencies of matched data. The number of useless steps (us2) at time t_0 will be compared with at time t_1, with statistical tests (parametric on not) for comparing the means of matched data (pre/post). A p-value of 0.05 will be considered significant.

(c) Evaluation of the effectiveness and usability of the TERENCE tool with the large scale evaluation. This is the most important objective of the evaluation. The research design is the crossover study design [7]. Children are subject to two different methods of reading: experimental method which is called TERENCE method, control method which is called TRADITIONAL method. The TERENCE method consists in reading stories using a computer system (see previous section). The TRADITIONAL method consists in reading stories using a paper-and-pencil approach. Children are randomly assigned to two groups called A and B. In first period of the experiment (I CYCLE), group A uses the TERENCE method and group B uses the TRADITIONAL method. During the second period of the experiment (II CYCLE), group A uses the TRADITIONAL method and group B uses the TERENCE method. In this experiment design, it is important to assume that the effect of the method of reading in each period is not affected by the method used in the previous period (carry-over effects). To minimise the possibility of the carry-over effects, we consider the "wash out" time between the periods in which the different methods of reading are used. Both groups are compared on the results and the effectiveness of method is determined. According to clinical practice, the two cycles shall last 2 months, and the wash out shall last 1 month. The primary goal is the evaluation of the level of the text understanding after the TERENCE use for effect of reading method and compare the level of the skills during the laboratory. For this endpoint, metrics, tests and expected results are the same of the small evaluation (pedagogical evaluation and usability-efficiency evaluation). For this goal, it is worth noting that during the stimulation of reading with the traditional method, measurements are performed manually by trained tutors. Data analysis is based on specific statistical models for crossover study.

4 Discussion

We here resume the advantages and the disadvantages from the different considered evidence based approaches. The advantages from the UCD perspective is that the EBM approach strongly reinforce the satisfaction analysis, and enables to measure how a usable software tool is able to improve certain abilities, in comparison with non-software methods (usually, in fact, UCD does not compare software tools with non-software tools). From the evidence-based viewpoint, UCD helps reducing the bias concerning the introduction of a software tool, since the tool is designed to be usable and with users put in the centre of the development. Furthermore, since UCD is flexible, it was easy to fit the usability study into the stricter EBM protocols. On the contrary, the stereotyped structure of the protocols usually written by EBM experts may represent a disadvantage, especially with respect to the cyclic and iterative nature of the UCD process, which instead requires more flexibility. The future work consists in applying the evaluation plan discussed above during the second half of the second year (for the small-scale evaluation), and for the third year of the project (for the large scale evaluation).

Acknowledgement. The authors' work was supported by the European Community Framework Programme 7, TERENCE project, contract number 257410, objective ICT-2009.4.2 TEL.

References

1. Bisiacchi, P., Cendron, M., Gugliotta, M., Tressoldi, P., Vio, C.: Neuropsychological battery for children 5-11 years old. Edizioni Erickson (2005)
2. Cain, K., Oakhill, J.V., Barnes, M.A., Bryant, P.E.: Comprehension Skill, Inference Making Ability and their Relation to Knowledge. Memory and Cognition 29, 850–859 (2001)
3. Camaioni, L., Ercolani, A.P., Lloyd, P.: Prova di Comunicazione Referenziale. Giunti O.S. (1995)
4. Cornoldi, C., Colpo, G.: Prove di Lettura MT-2 per la Scuola Primaria. Giunti O.S. (2011)
5. Dunn, L.M.: Peabody Picture Vocabulary Test - Revised - Italian version. Omega Edizioni (2000), Italian version and standardisation by: Stella, G., Pizzioli, C., Tressoldi, P.E.
6. Gulliksen, J., Gransson, B., Boivie, I., Blomkvist, S., Persson, J., Cajanger, A.: Key principles for user-centred systems design. Behavior and Information Technology 22(6), 397–409 (2003)
7. Jones, B., Kenward, M.: Design and analysis of cross-over trials. Monographs on statistics and applied probability. Chapman & Hall/CRC (2003), http://books.google.com/books?id=c
8. Marschark, M., Sapere, P., Convertino, C., Mayer, C.W.L., Sarchet, T.: Are Deaf Students' Reading Challenges Really About Reading? (in Press)
9. Raven, J.C.: Coloured Progressive Matrices. Giunti O.S. (2008)
10. Sackett, D., Rosenberg, W., Gray, J.A.M., Haynes, R.B., Richardson, W.: Evidence based medicine: what it is and what it isn't. British Medical Journal (1996)
11. Slegers, K., Gennari, R.: State of the Art of Methods for the User Analysis and Description of Context of Use. Tech. Rep. D1.1, TERENCE project (2011)
12. Tressoldi, P.E., Cornoldi, C.: Batteria per la valutazione della Scrittura e della Competenza Ortografica. Giunti O.S. (2000)
13. Yuill, N., Oakhill, J.: Effects of inference awareness training on poor reading comprehension. Applied Cognitive Psychology 2(33) (1988)

Assessing Connective Understanding
with Visual and Verbal Tasks

Magali Boureux, Barbara Arfé, Margherita Pasini, Barbara Carretti, Jane Oakhill, and Susan Sullivan

Abstract. The role of temporal and causal connectives is relevant in reading comprehension. Children with comprehension difficulties have problems in interpreting these connectives (e.g. Amidon, 1976; Feagans, 1980; Pyykkônen, Niemi and Järvikivi, 2003; Trosborg, 1982). The Adaptive Learning System (ALS) TERENCE aims to develop children's comprehension through the use of adaptive visual and verbal games. Within this framework, the purpose of this study was to assess connective comprehension with three visual and verbal tasks. Two hundred and eight English and Italian children participated in this study. The main results show that the use of pictures does not always support comprehension. Moreover, less skilled children perform better at simultaneous connective "while" compared to the temporal sequential connectives (before, after) and causal (because) ones.

1 Introduction

Reading comprehension is a daily activity that we often take for granted; however in reading and understanding a text several complex cognitive processes are engaged (such as language, reasoning and memory skills). Approximately 10% of young readers acquire age-appropriate word reading skills but do not develop commensurate reading comprehension ability [7], which is below the predicted level for their word reading ability and their chronological age. These children are

Magali Boureux · Margherita Pasini
Università degli studi di Verona

Barbara Arfé · Barbara Carretti
Università di Padova

Jane Oakhill · Susan Sullivan
University of Sussex

P. Vittorini et al. (Eds.): International Workshop on Evidence-Based TEL, AISC 152, pp. 19–26.
springerlink.com   © Springer-Verlag Berlin Heidelberg 2012

less likely than good comprehenders to integrate information in a mental representation of the text read [7]. In narrative texts, readers' ability depends on relating the narrated events to form a mental representation of their sequence. Readers use their knowledge of the language and their knowledge of the world to construct mental models of temporal and causal sequences of events narrated in a text. The main difficulty in this process is that language does not always encode events chronologically in a text. Some studies showed that children aged 7 to 12 sometimes fail to reach the correct interpretation of temporal sentences [2,12,19,24]. On the contrary, adults and more expert readers tend to store sequential events in chronological order [3,16,20,25,27], and do not have difficulties in achieving a coherent mental model of the event order whatever the order of the encoded events [10,13].

Developing children's ability to understand temporal and causal relations in stories is the goal of TERENCE, a EU funded Project aimed at designing an Adaptive Learning System (ALS) for poor readers and their educators. A first step is to examine children's comprehension of temporal and causal relations expressed by explicit connectives, like "before", "after", "while" and "because". Connectives are the main linguistic devices that help readers to establish relations be-tween the narrated events. They appear early in children's language production [23], but understanding of them is still developing in 10-year-olds [6], especially for connectives conveying complex cognitive relations, such as "while" [5]. We report the results of a study where the understanding of temporal and causal connectives was analyzed using tasks where the kind of information in support of the reader's reasoning was either visual (pictures) or verbal (text) (see material section).

Past studies of learning from text and pictures have shown that students learn better from text and pictures than from text alone [4,17]. In order to foster comprehension, information selected from the picture has to be integrated with information selected from text into a coherent mental representation [17,22,26]. Pictures in addition to text are especially suited for supporting cognitive functions that are not fostered by text alone [1,21]. Examples from visuo-spatial [15] or causal [14] reasoning show that such reasoning is easier when pictures support text. Some authors [9] argue that pictures can support critical psychological learning processes: they can support attention, help activate or build prior knowledge, minimize cognitive load and help to build mental models. On the basis of these observations, we predict that the comprehension of sentences with causal and temporal connectives should be easier when supported by pictures.

In our study, we compared the comprehension of causal and temporal connectives in a verbal context, given by a short narrative text, with comprehension of the same connectives in two tasks in which the context for interpreting the sentences was provided by pictures. Both sentence-pictures tasks respected the coherence between text and pictures, as suggested by Anglin, Vaez, & Cunningham, (2004) and Mayer (2005). In the first visual task, pictures helped the reader under-stand the events expressed by a sentence but not their (causal or temporal) relationship (Fig. 1) (tasks are described below). In this condition children's under-standing of the relation between the events depended only on reading and processing the connectives in each sentence. In comparison, in the second visual task one picture depicted a situation that was coherent with the temporal or causal relation (and connective) expressed by only one of three presented sentences. Children could understand both the events and their relationship looking at the picture, but they had to read (and understand) three sentences to choose the one that correctly described the situation in the picture (Fig. 2).

The performance of less skilled readers on the experimental tasks provides evidence concerning which tasks are more difficult, and the effects of visual aids on poor readers' performance. This information might provide indications for the development of the Artificial Intelligence Learning System (i.e. games).

2 Method

2.1 Participants

Sixty three English children (M=31; F=32) aged 7 to 11 years (M=9.03; SD=1.22) and 145 Italian children (M=76; F=69) aged 8 to 11 years (M=9.2; SD=.83) participated in the study. They attended schools situated in Sussex, UK (2 schools) and Veneto region in Italy (6 schools). All of the children spoke English or Italian fluently and had written parental consent to participate in the study. Data from those children with poor decoding abilities, or any known behavioral, emotional, or learning difficulties (provided by teacher reports) were excluded from the analyses reported in this paper.

Children were assessed with standardized reading tests: an adapted (listening) version of the Neale Analysis of Reading Ability—Revised British Edition [18] for English children, and, for Italian ones, "prove MT" - Revised Edition [11]. On the basis of their text comprehension score, children were classified as skilled comprehenders (SC) (SC English sample: N=35; Italian sample: N=102) or less skilled comprehenders (LSC English sample: N=28; Italian sample: N=43).

3 Materials

Three experimental tasks were set up to test the comprehension skills of temporal sequential (before, after), temporal simultaneous (while) and causal (because) connectives with different tasks, including pictures or not. Task 1 (T1) comprised 16 items. For each item, children had one sentence to read and 3 pictures portraying events expressed in the sentence, which were presented in a jumbled order. The child had to numerically order them (writing 1, 2 or 3 under each picture) according to the meaning of the sentence read (Fig. 1). In task 2 (T2), 21 items were presented. Each item included a picture that illustrated a situation or an event, and three sentences differing only in their connective. The picture was consistent with one of the causal or temporal relations expressed by the connectives, but not with the others. Children had to choose the sentence that best matched the picture (Fig. 2).

Poiché Alessandro aveva dimenticato lo zucchero e il detersivo, Isabella tornò al fare la spesa.

Fig. 1 Example of Task 1 item. Children had to fill the blank space with the right number to order the picture story.

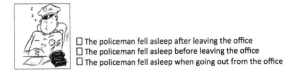

□ The policeman fell asleep after leaving the office
□ The policeman fell asleep before leaving the office
□ The policeman fell asleep when going out from the office

Fig. 2 Example of task 2 item. Children had to choose the correct sentence according to the picture.

Task 3 (T3) was a story of about 700 words (adapted from [8]). Twenty-four connectives of the story were substituted by groups of three connectives: one correct, in the verbal context of the text, and the other two wrong. Children had to choose the right one from the three, in order to restore the consistency of the story.

3.1 Design and Procedure

The experiment took place in school. Children were tested in small groups of five students (in the UK), or altogether in their own classroom (in Italy). The standardized reading comprehension tests (NARA and MT) were administered to all children to assess their reading skills. In order to avoid tiredness both Task 1 and Task 2 were split and half of each of the two tasks were administered at two different times. The order of presentation of the different parts was counterbalanced across participants. Task 3 was completed in a separate session.

3.2 Data Analysis

The results were analyzed for English and Italian children separately because of the differences between the languages and the reading tests used. Correlations between the scores on the reading tests and the three experimental tasks were run to test the validity of the tasks. We then conducted a mixed analysis of variance (ANOVA) to explore differences in correct responses (dependent variable), with task as the within factor (three levels: task 1, task 2, task 3), and skill group as the between factor (with two levels: SC and LSC). Since TERENCE is aimed at less skilled readers, a second analysis focused on this group was conducted. Two repeated measure ANOVAs were run, to verify differences in means of percentages of correct responses, one with the task as the within factor (T1, T2, T3), and one with the connective as the within factor (before-after; while; because), considering only the sample of LSC, separately for the Italian and English sample.

4 Results

The correlations between the comprehension scores and the tasks show differences between the two groups. For English children the correlation between reading comprehension scores (NARA) and Tasks 2 and 3 (picture-sentence match and verbal story tasks) was high (r=.54, p<.001 and r=.51, p<.001 respectively) and modest when Task 1 was considered (r=.27, p<.05). For Italian children, the results show that children's reading comprehension scores (MT tests) correlated with Task 1 and Task 3 (picture ordering task, r=.36, p<.001, and the verbal story task, r=.38, p<.001). No correlation was found between MT test and Task 2 performance.

The main effect of skill group was significant both for English ($F_{(1,35)}$=3.68, p<.001) and Italian children ($F_{(1,142)}$=45.1, p<.001), revealing that skilled comprehenders performed significantly better in all tasks. The analyses also revealed the main effect of task for English children ($F_{(2,70)}$=27.409, p<.001): Task 2 was easier than the other tasks. The same was found for Italian children ($F_{(2,284)}$=54.875, p<.001): the highest accuracy was for Task 2 and the lowest for Task 1. Also the interaction "Task" x "Skill Group" was significant for both English ($F_{(2,70)}$=4.519, p<.05) and Italian ($F_{(2,284)}$=5.279, p<.05) children. Task 3 was the best at differentiating between English SCs vs. LSCs. Task 2 was the one that differentiated the least between Italian SCs and LSCs: The higher accuracy of Task 2 yielded a ceiling effect for both groups (Tab.1).

The ANOVAs focused only on the less skilled readers showed that, in general, the performance was poorer in Task 1 than in Task 2. This difference was significant in both English ($F_{(2,34)}$=17.210, p<.001) and Italian ($F_{(2,84)}$=23.731, p<.001) children. The pattern of means was explored further with a series of paired sample t tests. For English children, performance on Task 3 was significantly poorer than performance on the other two tasks (p<.01). For Italian children, the poorest performance was on Task 1. However, Task 3 was performed less well than Task 2 (p<.005). These results confirm that the task supported by pictures (T2) is the most simple for children with comprehension difficulties. In contrast, the most complex tasks were T3 for English children and T1 for Italian ones. These results suggest that pictures per se do not support reading comprehension more than verbal information, but do so only when they convey information about the relationship between events represented in the text. This information does not necessarily need to explicitly represent the relationship between two or more events, but could be inferred from the use of appropriate pictures that trigger children's world knowledge (as in Fig. 2).

Table 1 Percentage of Correct Comprehension responses as a Function of Comprehension Skill and Type of Task for English (UK) and Italian (IT) children.

		Task 1	Task 2	Task 3
English	LSC	65 (5.6)	80.7 (3.5)	46.6 (5.3)
	SC	66.2 (5.5)	87.9 (3.4)	69.2 (5.1)
Italian	LSC	60.3 (2.9)	86.5 (2)	74.8 (1.9)
	SC	77.4 (1.9)	91.3 (1.3)	87.5 (1.3)

Standard Deviations are shown in parentheses.

The less skilled readers' comprehension of the connectives ("before-after", "while" and "because") revealed interesting results for both English and Italian readers. Indeed, we found that "while" was significantly better understood (UK mean=69.8, SD=4.2; It mean=83.5, SD=2.4) than "before-after" (p<.001) (UK mean=55.6, SD=4.1; It mean=66.7, SD=2.6) and "because" (p<.05) (UK mean=60.1, SD=3.8; It mean=71.6, SD=2.5) both by English ($F_{(2,54)}$=5.633, p<.01) and Italian ($F_{(2,84)}$=19.091, p<.001) less skilled comprehenders whereas we did not find differences in children's comprehension of "before-after" and "because" in either language. These results indicate that, in our Tasks, "while" relations were the easiest to understand for children with comprehension difficulties. Interestingly, the most complex connectives were the sequential ones "before" and "after".

5 Discussion

Correlational analyses showed that the experimental tasks requiring comprehension of temporal and causal connectives assess skills that are important for reading comprehension. However, whereas T2 discriminates well between English children with good and poor comprehension skills, it is less predictive of Italian children's comprehension because of ceiling effects. The differences between tasks results may be due to inherent characteristics of the two languages. Task 3 discriminates well between LSCs and SCs, mostly because, like the reading tests, it is a verbal task. This observation confirms the difficulty of children aged 7 to 11 years in interpreting connectives, as shown in [2,6,12,19,24].

As hypothesized on the basis of [4,17], pictures are useful for comprehension when they illustrate the situation described in the sentence (T2). In contrast, they do not support the comprehension of sentences when they do not illustrate the relationship between event and sentence (T1). In this case, accuracy is similar to that obtained in verbal task (T3). This finding shows that verbal tasks are not systematically more difficult than tasks supported by pictures. Pictures can make the task easier allowing inferences about the relation between events on the basis of the child's world knowledge, as in T2. When more complex cognitive processes are required, as in T1, pictures do not support verbal comprehension. In T1, readers could have difficulties not only when comprehending the sentence, but also when manipulating the pictures.

Interestingly, the sequential temporal connectives "before" and "after", and the causal one "because" were the most difficult, whereas "while" was the easiest. This pattern was found across tasks, and indicates that sequential events are probably more difficult to represent through language than simultaneous ones, contrary to that which was found in [5]. The difficulty for readers could be due to the fact they have to process the two sequential or causal events in the whole sentence context in order to understand which event is the first and which the second one. In contrast, "while" could be the most accurate because it mainly connects an event to a situation: its comprehension is linked more to reasoning processing and less to the sentence structure. Pictures can fully represent only one event at a time: thus, in the case of two sequential or causal events, the picture can represent only one of the two, and the reader has to infer the temporal or logical position of the one which is not represented. On the contrary, when representing "while" sentences, both pieces of information (event and situation) can be represented in the same picture. This observation seems to confirm the previous hypothesis that pictures help when they represent the relation between events.

6 Conclusions

The results of this research suggest some interesting implications concerning the comprehension of temporal and causal relations expressed by connectives in texts. Among them:

- verbal tasks are not systematically more complex than visual ones;
- pictures make reading comprehension easier when they allow inferences about the relation between events on the basis of the child's world knowledge, as in T2;
- the comprehension of "while" relations seems to be less sensitive to the kind of visual representation provided: a picture sequence or a single picture.

Acknowledgments. This research was supported by the European Commission through the Seventh Framework Programme for Research and Technological Development. Strategic Objective ICT- 2009.4.2 : ICT : Technology-enhanced learning for TERENCE project.

References

[1] Ainsworth, S.: DeFT: A conceptual framework for considering learning with multiple representations. Learning and Instruction 16, 183–198 (2006)
[2] Amidon, A.: Children's understanding of sentences with contingent relations: Why are temporal and conditional connectives so difficult. Journal of Experimental Child Psychology 22, 423–437 (1976)
[3] Andersson, A., Garrod, S.C., Sanford, A.J.: The accessibility of pronominal antecedents as a function of episode shifts in narrative text. Quarterly Journal of Experimental Psychology 35A, 427–440 (1983)

[4] Anglin, G.J., Vaez, H., Cunningham, K.L.: Visual Representations and Learning: The Role of Static and Animated Graphics. In: Jonassen, D.H. (ed.) Handbook of Research for Educational Communications and Technology, pp. 865–913. Simon & Schuster, NY (2004)

[5] Arfé, B., Di Mascio, T., Gennari, R.: Representations of Contemporaneous Events of a Story for Novice Readers. In: Magnani, L., Carnielli, W., Pizzi, C. (eds.) Model-Based Reasoning in Science and Technology. SCI, vol. 314, pp. 589–605. Springer, Heidelberg (2010)

[6] Cain, K., Nash, H.: The influence of connectives on young readers processing and comprehension of text. Journal of Educational Psychology 103, 429–441 (2011)

[7] Cain, K., Oakhill, J.: Profiles of children with specific reading comprehension difficulties. British Journal of Educational Psychology 76, 683–696 (2006)

[8] Cain, K., Patson, N., Andrews, L.: Age- and ability-related differences in young readers' use of conjunctions. Journal of Child Language 32, 877–892 (2005)

[9] Clark, R.C., Lyons, C.: Graphics for Learning: Proven Guidelines for Planning, Designing, and Evaluating Visuals in Training Materials. Pfeiffer, CA (2004)

[10] Claus, B., Kelter, S.: Comprehending narratives containing flashbacks: Evidence for temporally organized representations. Journal of Experimental Psychology: Learning, Memory, and Cognition 32, 1031–1044 (2006)

[11] Cornoldi, C., Colpo, G.: Nuove Prove di Lettura MT per la Scuola Media Inferiore. Organizzazioni Speciali, Firenze (1995)

[12] Feagans, L.: Children's understanding of some temporal terms denoting order, duration, and simultaneity. Journal of Psycholinguistic Research 9, 41–56 (1980)

[13] Gennari, S.P.: Temporal references and temporal relations in sentence comprehension. J. Exp. Psychol. Learn. Mem. Cogn. 30, 877–890 (2004)

[14] Hegarty, M.: Mental animation: Inferring motion from static displays of mechanical systems. J. Exp. Psychol. Learn. Mem. Cogn. 18, 1084–1102 (1992)

[15] Larkin, J.H., Simon, H.A.: Why a diagram is (sometimes) worth ten thousand words. Cognitive Science 11, 65–99 (1987)

[16] Mandler, J.M.: On the comprehension of temporal order. Language and Cognitive Processes 1, 309–320 (1986)

[17] Mayer, R.E.: Multimedia learning. Cambridge University Press, New York (2005)

[18] Neale, M.D.: The Neale analysis of reading ability—Revised British edition. NFER-Nelson, Windsor (1989)

[19] Pyykkönen, P., Niemi, J., Järvikivi, J.: Sentence structure, temporal order and linearity: Slow emergence of adult-like syntactic performance in Finnish. SKY Journal of Linguistic 16, 113–138 (2003)

[20] Radvansky, G.A., Zwaan, R.A., Federico, T., Franklin, N.: Retrieval from temporally organized situation models. J. Exp. Psychol. Learn. Mem. Cogn. 24, 1224–1237 (1998)

[21] Scaife, M., Rogers, Y.: External cognition: How do graphical representations work? International Journal of Human-Computer Studies 45, 185–213 (1996)

[22] Sless, D.: In Search of Semiotics. Groom Helm, London (1986)

[23] Spooren, W., Sanders, T.: The acquisition order of coherence relations: On cognitive complexity in discourse. Journal of Pragmatics 40(12), 2003–2026 (2008)

Concept Maps and Patterns for Designing Learning Scenarios Based on Digital-Ink Technologies

Félix Buendía-García and José Vte. Benlloch-Dualde

Abstract. The research on educational technologies is continuously growing and their application demands an additional effort to instructors. The current work describes the use of concept maps and patterns to provide a better support to instructors in the design of their technology-enhanced learning scenarios. The combination of both representation mechanisms has been applied in a Higher Education context based on the use of digital-ink technologies. The obtained outcomes reveal, first, the contribution of concept maps to represent technology-enhanced learning scenarios from a generic point of view and second, the feasible design of specific learning scenarios based on digital-ink patterns which have been generated from previous concept maps.

Keywords: learning scenario, concept maps, design patterns, digital-ink technologies.

1 Introduction

The research on educational technologies is continuously growing and their application demands an additional effort to instructors. There are multiple examples of technology-enhanced learning (TEL) settings either in classroom situations or on-line environments. However, instructors usually lack the necessary support to know in depth emerging technologies and how to effectively apply them in their courses. Therefore, specific approaches are required in order to "teach effectively with technology" [1] [2]. The current work describes the use of certain mechanisms which can contribute to provide a better support to instructors in the design of their technology-enhanced learning scenarios.

Félix Buendía-García · José Vte. Benlloch-Dualde
Department of Computer Engineering, School of Engineering in Computer Science,
Universitat Politècnica de Valencia, Camino de Vera s/n 46023, Valencia
e-mail: {fbuendia,jbenlloc}@disca.upv.es

P. Vittorini et al. (Eds.): International Workshop on Evidence-Based TEL, AISC 152, pp. 27–35.
springerlink.com © Springer-Verlag Berlin Heidelberg 2012

Design approaches can be based on different types of representation mechanisms ranging from simple narrative descriptions to more formal notations and specifications. These mechanisms fall within the field of Learning Design (LD) that deals with the need to guide and assist teachers in the preparation of effective learning scenarios [3]. In this context, the representation of instructional issues in visual formats is helping to improve the design of TEL scenarios [4]. This work proposes the use of concept maps and design patterns to represent information items in such scenarios taking advantage of their visual display possibilities [5].

To obtain evidences of the proposal, a special type of TEL scenarios has been selected, which is focused on digital-ink technologies. In particular, the flexibility of Tablet PCs and digital ink has the potential to achieve a wide range of educational goals and promote a more dynamic classroom environment [6], [7].

The current work introduces an approach to help practitioners in the design of content resources and teaching and learning tasks supported by these technologies combining the high level representation view provided by concept maps with design patterns which enable the design of learning scenarios adapted to certain technological settings.

The remainder of the work is structured as follows. The next section presents the use of concept maps and patterns to represent and design learning scenarios. The third section describes several case studies to show the application of these mechanisms in an educational TEL context based on digital ink technologies. Finally, some conclusions are remarked.

2 Designing Learning Scenarios

There have been multiple initiatives in the last years which have contributed to the modeling of LD issues. Computer Science and Software Engineering disciplines have promoted different notations and mechanisms in this context. Hypermedia models, ontology proposals, modeling languages, standard specifications or topic maps are some examples. These mechanisms provide several ways to represent learning issues in text or graphical format, using natural language or through a restricted vocabulary, and differing in their formalization level or abstraction degree. Moreover, there is a growing interest in visual instructional design languages [8] though they are mostly focused on a technological point of view disregarding pedagogical issues.

This work proposes the use of two types of complementary mechanisms to represent technology-based learning scenarios concerning visual display artifacts easy to understand by practitioners. The first one is based on concept maps that are graphical tools used by instructors to organize and represent knowledge. The second one deals with design patterns which have been usually applied in technological disciplines but they are also present in pedagogical and e-learning environments.

2.1 Concept-Map Based Design

Concept maps, proposed by Novak [8] a few decades ago, provide a very flexible structure allowing users to create meaningful relationships between key concepts which model the knowledge items to be represented. The semantic model is rather simple because it is composed by two main components: concepts that represent notions or ideas about a knowledge domain, and relationships which link such concepts. Fig 1 shows an example of conceptual map that displays some basic concepts that can be part of a learning scenario such as content resources, learning activities, interaction techniques or assessment procedures. These concepts can be related to derived concepts about the implementation of contents based on electronic presentations or multimedia resources, or learning activities such as open assignments or questionnaires.

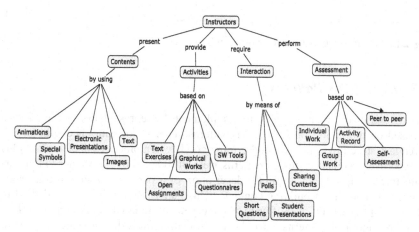

Fig. 1 Instructional concept map.

In a similar way, concept maps can be used to represent those notions and relationships involved in a technological-enhanced setting, for example, to model the knowledge about digital-ink technologies in a comprehensive manner. These technologies rely on LCD screens where a digitizer can capture the movement of a special pen and thus, allowing users to put data onto the screen in a natural way. Digital inking improves the chances for active learning activities allowing actions such as writing specific symbols, highlighting text areas, marking content items, or drawing diagrams. Fig 2 shows on its right side, part of a concept map representing some of these digital-ink capabilities. The concepts assigned to these capabilities can be linked to instructional concepts such as those modeled in Fig 2 (left side). This linking feature provides a useful connection between the requirements needed by instructors and the potential services provided by a technological-enhanced environment, such as one based on digital-ink technologies. In this way, the learning designer can be able to match instructional and technical issues by comparing the connected concepts and the involved relationships. This fact helps in the generation of guidelines to assist teachers in the instructional use of educational technologies [10].

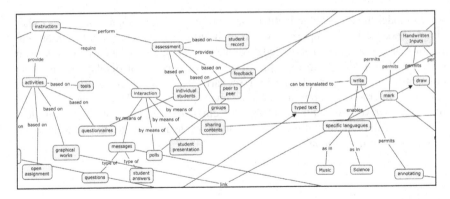

Fig. 2 Linking instructional and technological-based concepts.

2.2 Digital-Ink Patterns

The second mechanism proposed to complement the representation of TEL issues is based on the application of design patterns. The use of patterns can be considered as a structured method of describing good design practices in different fields of expertise. Originally, design patterns were introduced by Alexander [11] in architecture disciplines as "a careful description of a perennial solution to a recurring problem within a building context". This pattern notion has been adopted in other disciplines such as Software Engineering or Interaction designs. Furthermore, pedagogical patterns have been recognized as efficient mechanisms to document good practices in teaching including visual flow representations [12]. There are also design patterns which have been proposed in e-learning contexts "as conceptual tools to support educational design" [13]. These good practices can be extracted from technology-enhanced learning scenarios represented by concept maps such as those described in the previous subsection which linked instructional and technical items.

Therefore, patterns seem a powerful mechanism to allow instructors to design learning issues related to items such as theoretical contents or laboratory activities in a certain technology-based educational context. The approach proposed in this work is based on promoting a "guide rather than prescribe" philosophy to apply patterns, focused on small-scale learning experiences (previously based on concept map representations) and bounded to specific technology settings. Table 1 shows a summary of the language proposed to define patterns that fits the learning design philosophy aforementioned. This pattern language is mostly based in the original Alexandrian definition which is mainly narrative with some additional attributes and special features: i) the diagrammatic part is complemented with tags that specify particular concepts with a potential instructional purpose and ii) an extra field called Keywords that gathers some of the previous tags and other terms in order to characterize the learning scenario through the proposed pattern.

Table 1 Pattern language for learning design.

Name	Pattern identifier
Context	Description of the learning scenario in which the selected pattern is applied
Problem	Overview about the learning or instructional requirements to be faced
Discussion	Explanation to motivate the addressed problem and its justification
Solution	Description of the way to apply technologies to solve the addressed problem
Diagram	Sketch to represent the solution in a graphical display including descriptive tags
Relationships	Links to other patterns which could be useful in the learning scenario design
Keywords	Collection of terms which reference specific aspects of the learning scenario

Fig 3 shows an example of sketch that represents some of the actions associated to the use of digital ink technologies such as fixing a writing mistake or framing a content item. That visual representation is very easy to understand by non-computer literate users like most instructors, allowing them to recognize the digital-ink potentials in order to take advantage of these technologies in specific learning scenarios. Moreover, the inclusion of tags and keywords enable the pattern processing in an automatic or semiautomatic way, for example, through the use of ontology notations, and it represents a further step in the LD formalization.

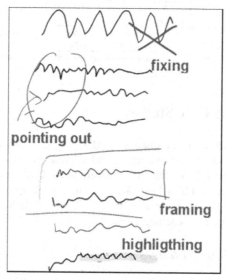

Fig. 3 Digital-ink sketch.

Another interesting feature of design patterns is the possibility to classify them into different categories to facilitate their further detection, definition and processing. Table 2 shows four categories used for classifying digital-ink patterns: *Content* managers, *Activity* facilitators, *Interaction* enablers and *Assessment* producers. Each category contains some examples including the pattern name and a short description.

Table 2 Digital-ink patterns.

Category	Name	Short description
Content manager	Light and shade	Some content items need to be clarified using an extra explanation or highlighted by means of visual artifacts.
	Focus of attention	There are items that require to be located, by signaling, underlining or framing certain information
	Half-baked	Some resources such as slide-based presentation can be completed on the fly by using freehand inputs to facilitate presentations.
Activity facilitator	Make connections	There are activities that require to link or set up relationships among their component items.
	Sharing efforts	Several students need to participate and collaborate to solve a problem, sharing and exchanging information.
	Filling blanks	Different activities can demand to introduce information on a previously prepared structure (text, table, diagram, map...)
Interaction enabler	Raise your question	Anonymous contributions can help students who are reluctant to ask in public (this pattern could be related with "Focus of attention").
	Post your opinion	Students can contribute with their point of view in a topic discussion.
	The audience answer	A poll mechanism can be used to gather the overall student preferences or the knowledge about a topic.
Assessment producer	The right option	A rapid answer to a closed set of questions (objective test) is required.
	Connection game	A learning activity based on matching options could be evaluated (this pattern could be related with "Make connections").
	Bad news	Instructor can signal or remark the corrections made in the student works (fixing common mistakes).

3 Implementing Case Studies

Some case studies have been implemented in order to show evidences of the combined application of concept maps and design patterns. In particular, they were essayed in several courses of undergraduate Computing courses to face instructional problems such as low participation and poor students' performance. These innovative strategies were supported by Hewlett Packard in the framework of the HP Technology for Teaching Grant Initiative [14].

3.1 Concept-Map Evidences

In order to get evidences, several workshops were carried out to present faculty the instructional potential of digital-ink technologies. In particular, attendants made use of a classroom equipped with Tablet PCs and presentation software called Classroom Presenter [7]. At the end of the workshop, once participants became familiar with technology, a questionnaire generated from the concept map shown in Fig 2 was distributed to ask them about the digital-ink possibilities in their context.

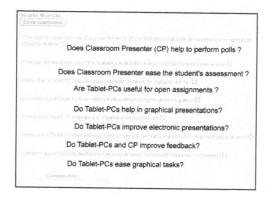

Fig. 4 Questionnaire about digital-ink instructional potential.

Fig 4 shows a form containing some of the questions generated from concept links connecting instructional and technological issues. Fig. 5 depicts the questionnaire results provided by faculty participants. For example, almost 80% agreed that these technologies could be useful to provide timely and rich feedback to students.

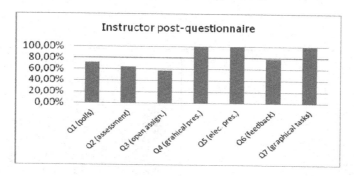

Fig. 5 Questionnaire results.

3.2 Pattern-Based Evidences

Previous workshop experiences were used to obtain a general view (concept map-based) of the instructor requirements accommodating them to the digital-ink potentials. This situation was exploited to generate digital-ink patterns which could be applied in specific learning scenarios. Fig 6 shows a diagrammatic description of a pattern sample called "Half-baked" that belongs to the Content manager category (see Table 2). It proposes designing presentations in such a way that slides carry only the key ideas allowing instructors to complete them on the fly using freehand inputs.

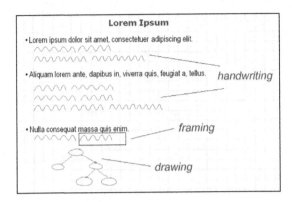

Fig. 6 Digital-ink pattern sketch.

Fig. 7 (a) shows an application of the aforementioned pattern on *Computer Technology*, a first-year course on the Computer Engineer degree. In this case, starting from a circuit diagram and a main sentence, instructors dynamically developed some arguments to support the concept explanation. In a similar manner, Fig 7 (b) shows an application example of a pattern called "Filling blanks" to carry out an in-class activity in the context of a *Data Structure and Algorithms* course.

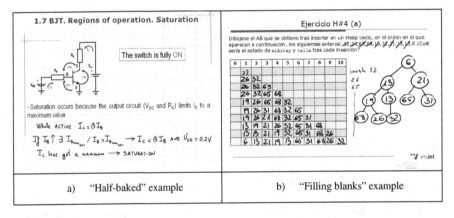

Fig. 7 Patterns implementation.

4 Conclusions

The current work has described the combined use of concept maps and patterns in the design of learning scenarios based in technology-enhanced settings. The proposed approach has taken advantage of concept maps in a first stage to represent instructional requirements and technological solutions. This stage provides a high-level design overview that permits to connect both sides from an abstract point of view. In a second stage, the pattern application has contributed to address more specific issues enabling the representation and design of particular learning scenarios adapted to certain technological settings. This combined approach has been applied in an educational technology-enhanced context based on digital ink technologies. The approach application has enabled the conceptualization of instructional and technological issues demonstrating the connection of requirements needed by instructors and the potential services provided by digital-ink technologies.

Acknowledgments. This work has been supported by the TEA project (PAID-UPV/2791) and ETSINF (Escuela Técnica Superior de Ingeniería Informática).

References

1. Dyrli, O.E., Kinnaman, D.E.: Teaching effectively with technology: What every Teacher needs to know about technology, Technology and Learning magazine (1995)
2. Wenglinsky, H.: Using technology wisely: the keys to success in schools. Teachers College Pres, New York (2005)
3. Conole, G., Weller, M.: Using learning design as a framework for supporting the design and reuse of OER. Journal of Interactive Media in Education 5 (2008)
4. Agostinho, S.: The use of visual learning design representation to document and communicate teaching ideas. In: Proceedings of ASCILITE, Sydney (2006)
5. Buendía, F., Díaz, P.A.: Framework for the Management of Digital Educational Contents. Educational Technology & Society 6(4), 48–59 (2003)
6. Huettel, L.G.: Transcending the traditional: Using Tablet PCs to enhance engineering and computer science instruction. In: 37th ASEE/IEEE Frontiers in Education Conference (2007)
7. Anderson, R., Davis, P., Linnell, N., Prince, C., Razmov, V., Videon, F.: Classroom Presenter: Enhancing Interactive Education with Digital Ink. Computer 40(9), 56–61 (2007)
8. Figl, K., Derntl, M., Caeiro Rodriguez, M., Botturi, L.: Cognitive effectiveness of visual instructional design languages. J. Vis. Lang. Comput. 21(6) (2010)
9. Novak, J.D.: Learning, creating, and using knowledge: Concept maps as facilitative tools in schools and corporations. Lawrence Erlbaum Associates, Mahwah (1998)
10. Buendia, F.: Supporting the Generation of Guidelines for Online Courses. Journal of e-Learning and Knowledge Society 7(3), 51–61 (2011)
11. Alexander, C., Ishikawa, S., Silverstein, M.: A pattern language: Towns, buildings, construction. Oxford University Press, Oxford (1977)
12. Hernández, D., Asensio, J.I., Dimitriadis, Y., Villasclaras, E.: Diagrams of learning flow patterns' solutions as visual representations of refinable IMS learning design templates. In: Handbook of Visual Languages for Instructional Design, pp. 395–413. IGI Group (2007)
13. Goodyear, P.: Educational design and networked learning: Patterns, pattern language and design practice. Australasian Journal of Education Technology 21(1), 82–101 (2005)
14. Higher Education HP Technology for Teaching Grant Initiative Recipients (2008)

Online Social Networks Impact in Secondary Education

Habib M. Fardoun, Daniyal M. Alghazzawi, Sebastián Romero López,
Victor M.R. Penichet, and Jose A. Gallud

Abstract. This paper presents and analyzes the potential uses and motivations of online social networks in education, with special emphasis on secondary education. First, we show several previous researches supporting the use of social networking as an educational tool and discuss Edmodo, an educative online social network. The work carried out during two academic years with senior students of primary and secondary schools is also analyzed. This research has allowed us to see the reality of social network use among young people and identify the challenges of its application to education environment.

Keywords: Online Social Networks, e-Learning, Secondary Education, Educative Web Tools, Collaborative Systems.

1 Introduction

Endless are the number of hours, news, energy that is now being used by the online social networks (hereafter OSN), adolescents being one of the main social groups that depend on these systems to communicate with their peer group and acquaintances. But how can it is possible to work with a tool that has many friends, messages, photos, videos! Which suppose too many distracting to the students, and would be contrary to achieve the objective sought. Some of the research

Habib M. Fardoun · Daniyal M. Alghazzawi
Information Systems Department, King Abdulaziz University (KAU)
Jeddah, Saudi Arabia
e-mail: {hfardoun,dghazzawi}@kau.edu.sa

Sebastián Romero López · Victor M.R. Penichet · Jose A. Gallud
ISE Research Group, University of Castilla-La Mancha
Campus universitario s/n, 02071 Albacete
e-mail: {sebastian.romero,victor.penichet,jose.gallud}@uclm.es

P. Vittorini et al. (Eds.): International Workshop on Evidence-Based TEL, AISC 152, pp. 37–45.
springerlink.com   © Springer-Verlag Berlin Heidelberg 2012

considered in this paper study is *Facebook* as a tool for college students with positive results. But for teenage students, is responsible enough to use the educational online social networks? Or to use these environments, the student, must be taken out from the OSNs that use commonly, and use others where there are no many elements that make the students lose concentration while they are performing their tasks.

The initial hypothesis to be validated: social networks can serve as an educational tool, acting as a motivator and enabler of social capital in education during adolescence (age range 12 to 18 years, with reference to the Spanish system, named as compulsory secondary education "E.S.O"). The paper is organized in four sections: Social networking in education, the analysis of one of the major educational OSNs, the field work which discusses the use of the OSNs by students, and it ends with conclusions and future work.

2   Social Networks within Education

This section introduces the basic concepts of the OSN, a series of research on its application in education, and the beginnings of the union of these two fields.

2.1   OnLine Social Networks

According to Danah Boyd [1], OSNs are Websites that give users a range of services based on Web technologies that allow individuals to: build a public or semi-public profile with relationships system, to have a list of other users with whom they share a connection, and finally, view and navigate through the list of users' connections with those who share a connection in the system. The shape and nomenclature of the connections listed above vary from one social network to another. What makes the OSNs unique, not because they allow users to meet others in the network, but because they make possible for users to manage and make visible their own social network. Normally connections on OSNs are between individuals who have "latent ties", Haythornthwaite [2], and that have some offline connection. In many OSN, users are not looking to expand their network of contacts (such as LinkedIn [3]), but they communicate with people they already knew prior to their entry into this OSN.

The most widespread and used features by OSNs users are: uploading and sharing photos and videos, comments on other profiles, friends and private messages between users. Users of these sites also share a number of documents and communicate with each other.

2.2   Computer Applications for Learning Support

Arguably, the first steps of social networks in education correspond to Moodle [4], at least in terms of the widespread use of the platform. Moodle is a project designed to support a social constructionist framework of education. It is distributed as free software (GNU). Moodle is copyrighted, but user can copy, use and modify Moodle if they agree to distribute the source code to others, without removing the original license and copyrights. The design and development of Moodle is based on an educational philosophy called "social constructionist pedagogy". Moodle can be considered as one of the first OSN focused on education, because it has one of the OSN main features.

Some basic questions that the Web Systems share with education are: Who are the students? What student's intentions and behaviours will be supported by the system? What devices will be used by students? E-Learning platforms solve these questions based on five different aspects: purpose, use, content, functionality and presentation. Based on the information taken from [5], in the following sections, EdModo [6] is discussed.

Moreover, in [5] we found comparison of the educational and technical aspects of the main electronic learning platforms: Blackboard Academic Suite 8.0, Claroline 1.8.1, Ecollege, WebStudy Course Management System, Atutor 1.5.4, Moodle 1.9, and JoomlaLMS. They compare various aspects like: productivity, communication, participation of students, administration, content development, licensing, and the required hardware and software. Highlighting after such detailed analysis the communication and motivation as key factors in the student learning process, therefore the student should not be or feel isolated. Finally the authors sort the platforms in two types: Those that are not attractive for most users, but at the same time they are fully developed and have most of the functionality needed by teachers and students. Those that are highly attractive, but do not provide a variety of services.

3 Edmodo Analyses

This section discusses the online social network tool Edmodo [6], an educative social network. The following we describe its implemented functionality, the non-permitted and weaknesses points. Based on the presented aspects in the sub-section 2.2, the purpose of this tool is the informal education, and to be used as an educational system and its contents are usually related to different subjects of the students.

3.1 Edmodo Main Features

In this sub-section we analyze the main features which are available on the Edmodo platform. It will be discussed some specific functions of communication, organization, file sharing and educational tasks.

The initial interface that the tool offers for teacher and students is very similar, but with some extra functionality in the teacher side, like: The first action offered by the tool for teachers is to create the class groups as it is required. Each group has a number of options that can be managed, if the user has a teacher role. The teacher can view the group members (students and teachers), He can archive and / or delete a group if it is necessary. From the public view, we may highlight that the teacher can decide the comments to be shared with people who are from outside of a specific group.

In terms of communication that is performed by using a board, it could be presented, by the teacher, to an entire group or as a private individual for each student. The teacher has four types of communication: (1) messages, (2) alerts, (3) assignment (or a task which can be rated later) and (4) vote. It is possible to add to each communication element: a file, a link (URL) or an existent item from the digital library. It has a section called "Who?" Where users can send messages in deferent ways to users: individual (private), students group, teachers and parents.

In the student side, the communication options are more limited than those of teacher, where they only have the option message, and they can only communicate in two ways: (1) with the entire class in public way or (2) in private way with the teacher.

Both teachers and students have access to a calendar, depending on the classes they teach, and the students to the classes to which they have joined where also they can view the deliverables or dates set by the teachers. These management features convert the Edmodo tool in a great tool for organizing and planning.

For storing and sharing files, there are two points of view in the Edmodo platform: The teacher view, where he can share folders with material for one or more of his classes. And the student view, where he has the option pack, with a space of 100 MB to store his files and / or class assignments.

Finally, from the user profile, other users can see (if they are connected to him): public activity, connections with teachers (if the user has a teacher role) their colleagues, besides seeing their school and classes that they manage or in which they participate.

3.2 Weaknesses and Not Allowed Features

This section describes the unpermitted or unimplemented functionalities within the platform and its weaknesses. Studding this information is helping us in the implementation process of our own OSN tool, which we called Tweacher. Tweacher is an OSN for educational purposes whose target audience is very similar to Edmodo. The weaknesses and the not allowed features we discover in Edmodo are:

- It has no option to send private messages between students, avoid forgetfulness, communication between students occurs globally.
- In the communication part, it has not implemented a chat tool. While, many other social networks (like Facebook, Tuenti, and Myspace) implement a kind of chat area for users.
- It does not work with photo albums and tags like other social networks. It works with generic file type, and do not allow the action of tagging them.
- It does not implement any kind of page in which the user can see the subject structure (index).
- Edmodo structure facilitates informal education; however, the order of the content of the courses and materials is not entirely clear.

– The functionality backpack, where students can save files that cannot be accessed by teachers, can be a weak point, since students could use it to save improper files.

4 Fieldwork: Using On-Line Social Networks by Students

This section presents a field study, which reflects the big use of the new technologies and social networks by high school students. Also, it highlights a set of advantages for their application in teaching. The field study is focused on Tuenti [7] and Facebook. First, we present the results obtained through an anonymous questionnaire given at three centres of the community *Castilla-La Mancha, Spain* (two secondary and one primary education centres). Next, we will discuss the main findings of this field study conducted during the last two academic years and applied over 425 students (381 secondary education and 64 of primary one).

4.1 Anonymous Questionnaires on the Use of Social Networks

We address first the results in secondary education talks (12 to 18 years, questionnaires to students from 1° to 4° of the E.S.O):

In the academic year 2009 / 2010, 282 questionnaires was carried out by students. The result was that, 88% of students use online social network Tuenti, taking into consideration that according to Spanish law it is illegal for children less than 14 years to use it. Well, the surveys were conducted at the beginning of 2010, so the students were born in 1996 and 1997 (which are 44.7% of respondents). Those are violating the terms of use of this tool, because they are minors, in particular 86.7% are registered on the social network Tuenti.

Given that 88% of the students use the Tuenti social network, from this percentage we extracted other interesting facts like: The average number of "friends" is 198.9 in the profile of each student. The average time spent connected to the Tuenti social network is 1 hour and a half per day. With regard to the social network Facebook, we have: 43.4% of respondents are registered in this social network.

We can highlight 88% of respondents who use the social network Tuenti as 61% of them are connected to the social network for more than one year and 70% have more than 140 "friends".

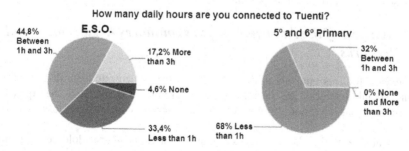

Fig. 1 Hours per Day using Tuenti, course 2010 / 2011

17.2% recognized to consume more than 3 hours use per day (Fig. 1 left.). While when we refer to harassment, the data is not worrisome since 3% have felt bullied at some point. With regard to Facebook, 63% of the students are registered on it, and the average number of friends is smaller to the average number of friends on Tuenti. The most alarming, and was not taken into consideration in the previous academic year's study, is whether the parents are concerned about what their children are doing using the OSNs (Fig. 2 left.). 55% respond positively to this question. This indicate that 45% of parents who are not interested and do not ask their children about the purpose behind using these new communication environments.

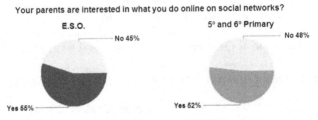

Fig. 2 Interest of parents for the use of ONSs, course 2.010/11

For primary education, the questionnaires were presented to 64 students in various lectures in 5° and 6° of the primary school, during the academic year 2011. We obtained the following results: 43.5% of students use Tuenti which suppose a high number, taking into consideration the prohibition that the Spanish authorities have with respect to this matter. On the other side, 52% of parents worry about what their children do in the OSN (Fig. 2 right), a slightly lower percentage compared to E.S.O parents. This slight difference would be probably because some of the children were not yet interested to register in OSNs. Statistics showed that 1.8% of respondents felt harassed using OSNs. And finally, we remark that the number of "friends" and hours of use per day is significantly lower than the results obtained in secondary education. Only 25% have more than 140 "friends" and 68% spend less than an hour online per day (Fig. 1 right).

We can say that, this trend of interaction through online social networks will continue in the future as the new generations make a high use of these networks because of being already an important part of their lives.

4.2 Analysis of the Anonymous Questionnaires Results on Social Networks

The reason for this analysis is that, in many cases where students do not have online social networks are because of the parents' controls, and the Spanish government that does not allow access to such services for citizens less than 14 years.

After analyzing this data, we can say that it is obvious where adolescents spend their time and what habits they have. Therefore, using this tool, as an educational, by students who know in depth and make use of their free time, can raise the motivation levels with respect to certain subjects. Although the learning curve of using the educational tools will be very quick, it would not reach the common social networks' level.

Could positive academic results be obtained through online social networks? The answer is yes, demonstrated by the results of the research presented next. In [8] research the impact of individual use of these online social networks from an educational point of view. This paper takes into consideration two processes of socialization such as: social acceptance and cultural adaptation, showing that online social networks in these processes help positively influence to academic outcomes. Thereby, it demonstrated the positive influence at the university level. So, obtaining positive results at the level of secondary education is one of the objectives of this work, considering that at this age, students need more control (from parents and teachers) than that at the college age.

4.3 Edmodo in a Real Environment

For this study we have worked with a group of 20 students in the "information technology and communication" subject of 1 bachelor's degree, and 38 students divided into two classes of 4° E.S.O in the "computer" subject during the months of February and March 2011. With respect to 1 bachelor's degree we highlight the following: We have dedicated and skilled jobs through the platform, there have been an informal communication between student-teacher and student-student. It was also proposed sharing of current information with the rest of the group. The study has been very successful. It was offered to teachers of 1 bachelor's degree the option of working with students using Edmodo. Only 14.3% seemed interested and responded positively (a possible weakness of this type of platform, teachers who feel unprepared or interested).

We conducted a survey with students obtaining the following data: Are OSNs interesting to use for educational purposes? 68.75% answered yes, 25% No and 6.25% indifferent. Would you like to work with Edmodo in other subjects? 66.7% said yes, 33.3% indifferent and 0% No. At the end, we highlight three data: 95% of students have participated in Edmodo, 70% actively and continuously (at least one publication a week), and 50% of students have used the platform in non-lecture hours.

The results obtained during 4° of the E.S.O haven't being as good as 1 bachelor degree, because its use by students was not as expected and their collaboration was proved to be too far below expectations. While bachelor students add news of every kind and discuss with them mates about them in an educate way, E.S.O. students just add contents to the wall of Edmodo when the professor have asked for it. Seeing the data of the sub-section 4.1 of this article the low participation in E.S.O. students was unexpected, ¿Which is the main problem? The difference in age is just one year, E.S.O. is compulsory education while bachelor isn't, differences between social groups and their behaviour as groups ¿could be them some explanations? Te unique way to resolve this question is by doing more field-research.

As a consequence of this field-research this year five professors more are using Edmodo at their classes, even with the results of E.S.O. students want new tools during their lessons. In this process we are learning how to teach using these new platforms that the research in e-learning gives us.

4.4 Others High Schools at a Click of Distance

In addition to what mentioned in the previous section Edmodo work between two classes of 1 bachelor degree in two different high schools (of Castilla-La Mancha). A new Edmodo group was created and it was supervised by two teachers in charge (one of each high school) established a set of rules of good use, otherwise teachers can use the option "reader" for students with bad behaviour. The group was of 41 students in the "information technology and communication" subject of 1 bachelor's degree. They were allowed to name the group to feel unit holders. And the main use was to share news and comment on technology among students of both high schools. The high school students who had been working longer with Edmodo were more active while their counterparts from the other school they were less (with certain exceptions).

Perhaps this extra motivation of being able to interact with peers from another part of his country, was what allowed us to obtain better results with bachelor students.

5 Conclusions and Future Work

This work is a study done for the implementation of a tool similar to Edmodo, whose name is Tweacher (Twitter + Teacher). In this study, we try to cover the weaknesses found in the previous tools (sub-section 3.2) and add extra functionality to improve the teaching-learning process through such tools.

With the presented data in section four, we can say that the use of OSNs in educational environment can be positive. And for its implementation as an educational tool, it takes time and especially for its involvement by the faculty.

A possible future improvement could be by establishing the use of counters to warn parents and teachers about the overuse of the tool by students. We did not analyze the parental control features available on Edmodo. It would be interesting to analyze and make a formal proposal on the possible ways of control that parents may have.

Acknowledgements. This research was partially supported by the project of the Ministry of Education and Science "CICYT TIN2008-06596-C02-0", the regional projects of the Regional Government of Castilla-La Mancha "PPII10-0300-4174" and "PII2C09-0185-1030". Thank you very much for teachers and students from the high schools of Castilla-La Mancha who participated in this research.

References

1. Boyd, D., Ellison, N.: Social Network Sites: Definition, History, and Scholarship. Journal of Computer-Mediated Communication 13 (1) (October 2007)
2. Haythornthwaite, C.: Strong, Weak, and Latent Ties and the Impact of New Media. The Information Society: An International Journal 18(5), 385–401 (2002), doi:10.1080/01972240290108195
3. LinkedIn red social on-line profesional, http://www.linkedin.com
4. Moodle 2.0.2. Recuperado el4 de Abril de (2011), http://moodle.org/
5. Elearnixml: towards a model-based approach for the development of e-learning systems. Tesis Doctoral de Habib Moussa Fardoun. Universidad de Castilla-La Mancha (2011)
6. EdModo red social on-line educativa, http://www.edmodo.com
7. Tuenti red social on-line, http://www.tuenti.com
8. Yu, A.Y., Tian, S.W., Vogel, D., Kwok, R.C.-W.: Can learning be virtually boosted? An investigation of online social networking impacts. Comput. Educ. 55(4), 1494–1503 (2010), doi:10.1016/j.compedu.2010.06.015

Learning to Read/Type a Second Language in a Chatbot Enhanced Environment

Giovanni De Gasperis and Niva Florio

Abstract. Evidence based design methodology can be applied to second language learning by introducing tools and methods based on human machine conversational agents such as restricted chatbots. General purpose chatbots have been used as English tutors, where the learner tries to maintain a generic conversation; on the contrary the proposed tools, obtained by an AIML chatbot generator, are aimed at having a restricted conversation with learners, specifically crafted for the second language training. In the first case study, the obtainable conversation is inspired by the exercises that typically are in a foreign language text book: the chatbot can corrects learners in real time whenever the learner produces incorrect sentences. On the other case, learners have to ask questions to a FAQ-chatbot about a fable that learners should have read to demonstrate the plot of the story has been understood.

Keywords: technology enhanced learning, conversational agents, AIML.

1 Introduction

For over 30 years educators have been be very interested in the use of computer-based technology in learning because of its inherent interactive capabilities and because it can also provide students with instant feedback to increase their autonomy [1]. In particular, concerning the use of this technology in learning a foreign language as second language, computers have been widely used as audio, video and

Giovanni De Gasperis
Dipartimento di Ingegneria e Scienze dell'Informazione, Matematica, Via G. Gronchi 18, 67100 L'Aquila
e-mail: giovanni.degasperis@univaq.it

Niva Florio
Scuola del Dottorato di Ricerca in Informatica, Dipartimento di Ingegneria e Scienze dell'Informazione, Matematica, Via Vetoio 1, 67100 L'Aquila
e-mail: niva.florio@univaq.it

P. Vittorini et al. (Eds.): International Workshop on Evidence-Based TEL, AISC 152, pp. 47–56.
springerlink.com   © Springer-Verlag Berlin Heidelberg 2012

graphics support. The birth and expansion of internet have contributed to increase the tools available for computer-based technology learning, for example providing students with 3D virtual environment where students communicate in English about everyday life topics as they are in a foreign country [2, 3] or talking with a English speaker conversational agent (e.g. [4]). Chatbots as tools to improve the learning of a foreign language are even more numerous on the web, but general purpose conversational agent as A.L.I.C.E. [5] are not very suitable for this purpose because chatting with them is based on the assumption that students already have a good knowledge of English as foreign language (see Sec. 2). The need to base TEL (Technology Enhanced Learning) techniques on pedagogical theories and design principles has become increasingly strong [1] and Evidence Based Design can improve the implementation and the use of TEL tools. The chatbot as an evidence based design tool could be applied to learning in general, given its efficacy to the discipline of interest. In this work the application of chatbots is studied in the context of second language learning were specially crafted and restricted chatbots are generated to fulfill learning goals posed by the educator.

2 Background

The use of computer, internet and intelligent agents related to TEL environment has many research contributions [6, 7, 8, 9, 10]. More recently, the use of conversational agents in TEL is being studied [11, 12] because the use of chatbots increases the motivation of learners to study [12, 13]. General purpose conversational agents have been used as tutorbots [14, 4], learning companions [15], pedagogical agents [16] and helpers [9, 17]. More specifically chatbots are seen as language learning tools [18], and some of them are developed for this purpose [4, 19, 20]

In this paper we proposed two case studies where restricted chatbots help second language learners. General purpose chatbots used as English tutors where the learner tries to maintain a generic conversation interacting with them via a textual interface it may not be appropriate at the first stage of learning the second language; on the contrary the tools we propose here are not aimed to have a generic conversation with learners. In the first case the chatbot would reproduce the exercises that typically are in a foreign language text book where the aim is the correct spelling of words. This kind of chatbot would not implement just a spell corrector, but could be enriched with knowledge about the words so that it can sustain a small conversation using them. The second case puts the learner in the situation to ask questions to an answering FAQ-chatbot expert about a small children fable to demonstrate she/he have read it and she/he understood the plot of the story.

As some other conversational agents [4, 19], the chatbots presented here are implemented writing AIML (Artificial Intelligence Markup Language [21]) knowledge-bases, interpreted by an online service, like Pandorabots [27]. A knowledge-base of this kind of chatter-conversational agent is made of pairs of textual/lexical (pattern; template), which can be linked together semantically and/or recursively by means of SRAI connections [21, 5]. The underlying model can be related to case-based reasoning semantic networks, as described in [22], based on textual pattern-matching algorithms [5].

As researchers assert in [23] and [24], a design-based methodology is important for the development of TEL strategies because a difference exists between what TEL environments are and what they would be, otherwise the resulting tools could not be effective to reach the desired learning goals . Evidence based design could be very useful for this purpose because, as Wang and Hannafin say in [23], *"designs are evidence based"*, so that we could read it in order to base on actual achievement of proper learning goals.

Evidence based design was developed [25] in a health care context where living environment, procedures and social interaction with patients should contribute to the well being and shorten the path to cure, or at least giving the patient a less stressful condition. Similarly in a context of TEL systems the context is made of software tools, graphical interfaces and reasoning algorithms that interact with the learner. The evidence based design in a TEL context should take care of the actual enhancement of learning, should measure actual progress of the learner cognitive profile and should motivate the learner to conquer better levels of understanding.

3 Chatbots as Language Tools

It has been demonstrated [26] that a restricted chatbot capable to chat in any language in a specific knowledge domain can be generated by starting of a simple input data set, made of:

1. a **FAQ** file F, frequently asked questions, is a free text file composed of several units of FAQ-knowledge:

   ```
   Q <the question phrase> |
     {Q <alternative version of the question>}
   A <the answer phrase> |
     {A <alternative version of the answer>}
   ```

   for as many units as needed to cover the restricted knowledge domain.
2. a **glossary** file G, where important keywords and or multi-word expressions will be listed with their free text definition:

   ```
   G <the glossary item>
   D <the glossary definition>
   ```

3. a **keywords** file K, just listed one each text line
4. a **multiwords** file M, just listed as many as needed on each text line
5. a **stop list** file S, listed all of the non meaningful words, like articles, prepositions, adjectives, adverbs and other forms.

Text files $< F, G, K, M, S >$ can be directly typed by the language educator using a simple text editor; this set makes the input of the PyGenBot software package, as described in [26], the educator obtains as output the set of ready-to-go AIML files, i.e. linguistic knowledge base files; then the AIML set can be fed to an online interpreter such as Pandorabots [27]. In this process the language educator does not have to know how to craft AIML categories because the AIML generation process is all internal into the PyGenBot package. The final result of publication of the AIML chatbot files on Pandorabots is a URL that can be used to exchange text with the generated chatbot; this URL can be later embedded in HTML code in any kind of Learning Management System, like the open source software package Moodle [28].

4 Proposed Evidence Based Methods

Two case studies are described, based on evidence based design principles, using a general purpose chatbot generator in second language learning projects:

- **Learning by typing**: how to generate a chatbot designed to correct misspelled words in typing session between a second language student and a properly generated AIML chatbot. Like in a typical language unit, a student should learn to read and type a specific set of words, with the proper spelling in a fixed case set of sentences/questions. S/he may also appreciate to use a small on-line dictionary available during the chat session just by typing the single word and having back its translation in the object language from the chatbot itself. This case study is shown in Section 5.
- **Learning by asking questions**: in a context were the language educator would like to improve the text comprehension of the learners, it is possible to generate a chatbot aimed to be an expert about a simple story or children fable. This case study is shown in Section 6.

In order to verify the evidence toward the efficacy of the proposed methods, further work should be carried out applying them in a context of language school training and evaluating their results. In general some learning metering parameters can still be defined and they will be described where applicable.

5 Case Study of a Chatbot for Word Spelling Improvement

The case study is designed to be simple in order to keep small size of the data set. Let us suppose to have an Italian speaking student who wants to read/type English as a second language. The data set includes a FAQ file made of 3 questions and their respective answers; each question includes 4 variants of the text varying in function of typing errors usually made by Italian students. The answers are the proper correct spelling sentences in the second language. In theory a new chatbot can be generated for each language lesson unit, according to the planned language learning path, using a different set of $< F, G, K, M, S >$ files. In this case a candidate for a good learning metering parameter is the number of steps the learner takes to type the word correctly.

As an example, the following sections describe the set of input data from which the chatterbot can be generated using PyGenBot.

FAQ File

It contains the frequently asked questions in the format described in Section 3, but now the question should not be intended as a real question, but more of a correspondence between different text variant, with typing errors, of the correct phrase, that is the text of the answer.

Table 1 The Frequently Asked Question input set about the spelling corrector in English

Q: God morning; Q: Gud morning; Q: Good mornin; Q: Good moning; A: Good morning
Q: Nice to met you; Q: Nice to meet yu; Q: Nic to meet you; Q: Naice to meet you; A: Nice to meet you
Q: Uere do you come from? Q: Uer do you come from? Q: Where do you cam from? Q: Were do you come from? Q: Where do iu come from? Q: Where do yu come from? A: Where do you come from?

Glossary File

The glossary file contains a small set of glossary items in the format described in Section 3. It is used to implement a small and simple online translation dictionary available to the language learner. For each item the second language word is placed after the G letter, and the right translation is placed into the second part beginning with a D letter. This list of words should be carefully designed by the educator with the aim to increase the vocabulary of the learner.

Table 2 The Glossary input set of the spell corrector in English used as a vocabulary.

G: **Good morning***; D: Buongiorno*
G: **Good evening***; D: Buonasera*
G: **Good afternoon***; D: Buon pomeriggio*
G: **Good night***; D: Buonanotte*
G: **Goodbye***; D Arrivederci*

The keywords file is not needed if the chatbot is kept simple.

Multiwords

The list of multiwords is needed to fix the correct spelling of each sentence, now corresponding to the answers. In this way the multiwords used in the exercise will not be split into its component words when generating the AIML category by PyGenBot.

```
good morning, meet you, come from
```

6   Case Study of a Fable Expert FAQ-Chatbot

The case study is designed to be simple in order to keep small the size of data set. Let us suppose to have an English speaking student who would improve her/his comprehension of the Italian text as a second language. Given a portion of a the Aesop's fable entitled "*Il Cerbiatto e Il Cervo*", the educator can write a set of 13 questions/answers with mostly one text versions of each question, or multiple version where needed. As an example, the following is the full set of input data from which the AIML FAQ-chatterbot is generated using PyGenBot.

6.1   Aesop's Fable

The Italian text of a portion of Aesop's fable "*Il cerbiatto e il cervo*":

> "*Durante un bel mattino di fine inverno, mentre il grande cervo brucava tranquillo le foglie dei cespugli più bassi in compagnia dell'inseparabile figliolo, un possente ruggito squarció il silenzio della foresta. Era un leone! Il cerbiatto sconcertato osservó il suo babbo e, con enorme stupore scoprí che questi tremava come un fuscello al vento.*"

FAQ File
The FAQ free text file shown in Table 3 contains the frequently asked questions list about the Aesop's fable in the format described in Section 3. The questions are properly crafted by the educator in order to check if the learner is improving her/his comprehension of the text and the plot of the story. The answer sentence for each question should be kept simple, but complete to describe fully a concept.

Glossary File
The glossary text file shown in Table 4 contains the linguistic items in the format described in 3. It is used to implement an online small glossary available to the language student that would like to know more about an unknown word.

Keywords
The keywords file is useful to list the main concepts/entity/action of the story; it will optimize pattern matching of the user text during the chat session.

```
storia, brucava, mattino, inverno, cervo, silenzio,
osservo', sentiva, vide, tremava, leone, foresta,
babbo, figlio, ruggito, compagnia, cerbiatto,
frusciello, vento
```

Multiwords
The multiwords list is needed to fix words together that are related to a single semantic or lexical element, that is present in the fable; this list also contributes to optimize the internal pattern matching algorithm of the AIML interpreter when using the files generated by PyGenBot.

Table 3 The Frequently Asked Question input set about the Aesop's fable in Italian

Q: Dov'è ambientata la storia? A: la storia è ambientata in una foresta

Q: Chi brucava? A: Il grande cervo era intento a brucare

Q: Cosa brucava il grande cervo? Q: Cosa mangiava il grande cervo? A: Il grande cervo brucava le foglie dei cespugli più bassi

Q: Chi c'era con il grande cervo? A: Il grande cervo era in compagnia dell'inseparabile figliolo

Q: Che cosa squarció il silenzio? A: Un possente ruggito squarció il silenzio della foresta

Q: Verso chi volse lo sguardo il cerbiatto? Q: Chi osservó il cerbiatto? A: Il cerbiatto osservo' il grande cervo suo babbo

Q: Come si sentí il cerbiatto? A: Il cerbiatto rimase sconcertato

Q: Cosa vide il cerbiatto? A: Il cerbiatto vide che il babbo tremava

Q: Come tremava il babbo? A: Il babbo tremava come un fuscello al vento

Q: Chi squarció il silenzio nella foresta? A: Il leone squarció il silenzio della foresta

Q: Prima del ruggito cosa stavano facendo? A: Prima del ruggito il babbo brucava in compagnia del cerbiatto

Q: Quando brucavano i due cervi? A: Un bel mattino di fine inverno

Q: Con chi era in compagnia il cerbiatto? A: Il cerbiatto era in compagnia del suo babbo, il grande cervo

Q: Perché il babbo tremava? Q: Perché il cervo tremava come un frusciello al vento? A: Il babbo tremava perché il leone aveva ruggito

Q: Chi è l'inseparabile figliolo? A: Il cerbiatto è l'inseparabile figliolo del grande cervo

Table 4 The Glossary input set in Italian

*G: **inverno**; D: la stagione più fredda dell'anno*

*G: **cervo**; D grande animale con corna ramificate*

*G: **leone**; D grande felino, re della foresta, con una grande criniera*

*G: **vento**; D l'aria che si muove all'aperto*

tra le foglie degli alberi

*G: **foresta**; D insieme di tanti alberi di molte specie distribuiti su un vasto territorio*

*G: **babbo**; D padre, genitore maschio*

*G: **figlio**; D essere vivente discendente da due genitori*

```
grande cervo, stavano facendo, fine inverno
inseparabile figliolo, aveva ruggito
```

Stoplist
The following is an example of a short stop list of words that should not be counted for pattern matching, that usually are common words and do not carry semantic meaning.

```
e,a,nella,il,lo,la,l,al,un,una,del,
era,come,si,in,perch\'e,chi,che,cosa
```

Table 5 Interaction log of the conversation with the Italian training FAQ-chabot. U is the student, B if the chabot

U: chi brucava? B: Il grande cervo era intento a brucare
U: cosa mangiava il grande cervo? B: Il grande cervo brucava le foglie dei cespugli piú bassi
U: cosa mangiava? B: (unknown terms)
U: cosa brucava? B: (unknown terms)
U: cosa brucava il grande cervo? B: Il grande cervo brucava le foglie dei cespugli piú bassi
U: cosa brucava il cervo? B: Il grande cervo brucava le foglie dei cespugli piú bassi
U: chi c' era con il grande cervo? B: Il grande cervo era in compagnia dell'inseparabile figliolo
U: che cosa squarció il silenzio? B Il costrutto che hai usato risulta ambiguo per la mia base di conoscenza. Avrei 2 risposte diverse da darti, ma non sono in grado di selezionare quella giusta
U: che cosa squarció il silenzio della foresta? B: Il leone squarció il silenzio della foresta
U: Verso chi volse lo sguardo il cerbiatto? B: Il cerbiatto osservó il grande cervo suo babbo
U: Chi osservó il cerbiatto? B: Il cerbiatto osservó il grande cervo suo babbo
U: come si sentí il cerbiatto? B: Il cerbiatto rimase sconcertato
U: cosa è un cervo? B: grande animale con corna ramificate
U: cervo B: grande animale con corna ramificate
U: cosa vide il cerbiatto? B: Il cerbiatto vide che il babbo tremava
U: che vide il cerbiatto? B: Il cerbiatto vide che il babbo tremava
U: come tremava il babbo? B: Il babbo tremava come un fuscello al vento
U: cosa stavano facendo prima del ruggito? B: Prima del ruggito il babbo brucava in compagnia del cerbiatto
U: quando brucavano i due cervi? B: Un bel mattino di fine inverno
U: con chi era in compagnia? B: Il cerbiatto era in compagnia del suo babbo, il grande cervo
U: perché il babbo tremava? B: Il babbo tremava perché il leone aveva ruggito
U: perché il babbo tremava B: (unknown terms)
U: perché il cervo tremava come un fruscello? B: Il babbo tremava perché il leone aveva ruggito
U: chi è l'inseparabile figliolo? B: il cerbiatto è l'inseparabile figliolo del grande cervo
U: chi è Carlo magno? B: "Carlo Magno": termini non presenti nel glossario.

7 Conclusions

Restricted chatbot used as an evidence based learning tool can provide ways to assess the progression of learning to a second language student. In the case of the spelling corrector chatbot is possible to measure how the student improves her/his language skills by counting the number of steps necessary to achieved correct spelling. In the case of the fable expert FAQ-chatbot the learner increase her/his motivation to understand the semantic relations behind the story by interacting with chatbot; the textual interaction between learner and the FAQ-chatbot is shown in table 5; the student has to deeply understand the story to have a proficient interaction with the chatbot. The educator can then judge the learning improvement of the learner by analyzing the log file produced during the conversation with the fable expert FAQ-chatbot. Unanswered questions are the result of the limitation of the input data set, but they can also be considered by the educator as a way to judge how skilled the learner has become in the second language. The educator can also include the new unanswered questions used by the learner to produce a second stage of difficulty level, generating a new fable expert FAQ-chatbot, starting a recursive process that will eventually increase the skill level of the learner and the FAQ-chatbot itself. The fable expert FAQ-chatbot generated with the data set described is this paper is available online at Pandorabots [29]. The research of this paper was supported by the European Communitys Seventh Framework Programme FP7/2007-2013 under the TERENCE grant agreement n. 257410.

References

1. Liu, M., Moore, Z., Graham, L., Lee, S.: A Look at the Research on Computer-Based Technology Use in Second Language Learning: A Review of the Literature from 1990-2000. Journal of Research on Technology in Education 34(3) (2002)
2. Jiang, X., Liu, C., Chen, L.: Implementation of a project-based 3D virtual learning environment for English language learning. In: 2nd International Conference on Education Technology and Computer (ICETC), vol. 3, pp. 281–284 (2010)
3. Avalon (2009), http://www.avalonlearning.eu/ (accesed November 10, 2011)
4. An English Tutor (2009), http://www.rong-chang.com/tutor.htm (accessed November 17, 2011)
5. Wallace, R.S.: The Anatomy of A.L.I.C.E. In: Parsing the Turing test. Part III, pp. 181–210. Springer, Netherlands (2009)
6. Casamayor, A., Amandi, A., Campo, M.: Intelligent assistance for teachers in collaborative e-learning environments. Computers and Education 53(4), 1147–1154 (2000)
7. Tian, J.: Agent-based on-line collaborative learning environment. In: Workshop on Intelligent Information Technology Application (2007)
8. Bennett, C.L., Pilkington, R.M.: Using a virtual environment in higher education to support indipendent and collaborative learning. IEEE (2001)
9. Feng, D., Shaw, E., Kim, J., Hovy, E.: An intelligent Discussion-bot for answering student queries in threaded discussions. In: Proceeding of the International Conference on Intelligent User Interfaces IUI (2006)
10. Konstantidinis, A., Tsiatsos, T.: Selecting a networked virtual environment platform and the design of a collaborative e-learning environment. In: 22nd International Conference on Advanced Information Networking and Applications Workshop (2008)
11. Kerly, A., Hall, P., Bull, S.: Bringing chatbots into education: Towards natural language negotiation of open learner models. Know.-Based Syst. (2007), doi: 10.1016/j.knosys.2006.11.014
12. Kerly, A., Ellis, R., Bull, S.: Conversational Agents in E-Learning. In: Applications and Innovations in Intelligent Systems, pp. 169–182 (2009)
13. Chih-Yueh, C., Tak-Wai, C., Chi-Jen, L.: Redefining the learning companion: the past, present, and future of educational agents. Computers and Education 40(3), 255–269 (2003)
14. De Pietro, O., Frontera, G.: TutorBot: An Application AIML-based for Web-Learning. In: Advanced Technology for Learning (2005), doi: 10.2316/Journal.208.2005.1.208-0835

15. Eynon, R., Davies, C., Wilks, Y.: The Learning Companion: an Embodied Conversational Agent for Learning. In: Proceedings of the WebSci 2009: Society On-Line (2009)
16. Veletsianos, G., Heller, R., Overmyer, S., Procter, M.: Conversational agents in virtual worlds: Bridging disciplines. British Journal of Educational Technology (2010), doi: 10.1111/j.1467-8535.2009.01027.x
17. Vieira, A.C., Teixeria, L., Timteo, A., Tedesco, P., Barros, F.: Analyzing online collaborative dialogues: The OXEnTCHŁ-Chat. In: 7th International Conference on Proceedings of the Intelligent Tutoring Systems (2004)
18. Fryer, L., Carpenter, R.: Bots as Language Learning Tools. Language Learning and Technology 10(3), 8–14 (2006)
19. An English Tutor for non native speaker (2011), http://www.speakglobal.co.jp/member/bots/meg_trial.php (accessed November 17, 2011)
20. Ji, J.: The study of the application of a web-based chatbot system on the teaching of foreign languages. In: Proceedings of the 15th Annual Conference of the Society for Information Technology and Teacher Education, pp. 1201–1207 (2004)
21. Wallace, R.: AIML 1.0.1 reference (2005), http://www.alicebot.org/TR/2005/WD-aiml (accessed November 16, 2011)
22. Smid, K., Pandzic, I.G.: Conversational virtual character for the web. In: Proceedings of Computer Animation 2002, Geneva, Switzerland, pp. 240–247 (2002)
23. Wang, F., Hannafin, M.J.: Design-based research and technology-enhanced learning environments. Educational Technology Research and Development 53(4), 5–23 (2005)
24. Hannafin, M.J., Hannafin, K.M., Land, S.M., Oliver, K.: Grounded practice and the design of constructivist learning environments. Educational Technology Research and Development 45(3), 101–117 (1997)
25. Eddy, D.M.: Evidence-based medicine: a unified approach. Health Affairs (Project Hope) 24(1), 9–17 (2005)
26. De Gasperis, G.: Building an AIML Chatter Bot Knowledge-Base Starting from a FAQ and a Glossary. JE-LKS, Journal of e-Learning and Knowledge Society 2, 79–88 (2010)
27. Pandorabots: A Multilingual Chatbot Hosting Service, http://www.pandorabots.com/botmaster/en/home (accessed November 18, 2011)
28. A Learning Management System, http://moodle.org/ (accessed November 13, 2011)
29. The Aesop's novel expert FAQ-chatbot generated for this work, http://www.pandorabots.com/pandora/talk?botid=93a750848e3429d2 (accessed November 22, 2011)

Learning about Literature on the Web in a German School

Eva Holdack-Janssen and Ivana Marenzi

Abstract. This paper discusses new ways of using web resources for teaching German literature at school using the LearnWeb2.0 system. The group project-work focused on teaching the literary work of Wolfgang Borchert, a German writer of the postwar era, to two groups of pupils in a German high school. Even thought the project time was limited to three weeks most students selected good quality material and created very good presentations.

Keywords: Web2.0, Media Education and cooperative learning, Secondary Education, Computer-supported cooperative learning.

1 Introduction: Media Education and Cooperative Learning

Today, educators largely agree that printed media is not the dominant source of communication any longer and must be seen in relation to other inter-media reference systems. An integration of the didactics of literature and media is therefore desirable [9]. Pupils must attain a critical-reflexive media literacy in order to be able to resist the manipulation by the media and instead use them to articulate own interests [7, 8].

Web 2.0 tools and social software applications such as *YouTube* or *Flickr* provide new means to connect people to digital knowledge repositories and to other people in order to share ideas, collaboratively create new forms of content,

Eva Holdack-Janssen
Leinbizschule, Hanover
Röntgenstr. 8, 30163 Hanover, Germany
e-mail: e.holdackjanssen@googlemail.com

Ivana Marenzi
L3S Research Center, University of Hanover
Appelstrasse 4, 30167, Hanover, Germany
e-mail: marenzi@L3S.de

P. Vittorini et al. (Eds.): International Workshop on Evidence-Based TEL, AISC 152, pp. 57–65.
springerlink.com   © Springer-Verlag Berlin Heidelberg 2012

get effective support, and learn with and from peers. The success of Web 2.0 and platforms demonstrates that people are willing to share resources with other people in their private life. This possibility has not been adopted in education so far, even though the plethora of media and resources on the Web can be very effective in education and offer multi-perspective approaches to teaching different subjects. Instead of a pure acquisition of knowledge, the Web 2.0 can foster inductive, integrative and synthetic learning through actively structuring collaborative and creative activities. Learning on the Web must deal with personal identification and decision processes. Open and cooperative learning arrangements are useful in the task management and enable a high level of task-related activity [6] but good tools that can support this kind of interaction are still under development and evaluation. In Germany a few experiments have been carried out in schools to test a variety of educational software. In particular, Lower Saxony's core curriculum for high and middle-high school education in the subject German demands the expertise of "dealing with the media" including the Internet.

In our paper we present a case study in which students of two high school classes collaboratively searched in the Internet, collected and discussed materials about post-war literature, and created a final presentation. In this context, we describe the course design (the teacher´s methodological and didactic approach in order to develop a critical attitude to the resource) as well as the design and the evaluation of LearnWeb2.0 (a collaborative environment that can support students in searching, annotating and discussing internet resources).

2 Case Study

In our research project we decided to introduce media literacy at the Leibnizschule in Hanover as part of a series of German lessons about post-war literature. The case study involved 33 pupils from 8th grade (14 years old) and 19 from 12th grade (18 years old) collecting and discussing materials from the Web.

The aim of this project was to teach students to search in various Internet sources for new information related to the short story "An diesem Dienstag" written by Wolfgang Borchert [3, 4]. To select critically suitable material and to discuss and organize the results in final presentations, we used the *LearnWeb2.0* system, shortly described in the following section.

Since the available information on the Web is not always reliable and adequate for the classroom, a further objective was to support the students to carry out a critical and reflective use of online resources.

2.1 The LearnWeb2.0 system

LearnWeb2.0 is an online platform for sharing Web 2.0 resources, created within the TENCompetence European project and further developed at the L3S Research Center in Hanover. It provides several features designed to support teachers and students in collaborative group search and annotation of resources from different Web 2.0 services including *YouTube, SlideShare,* and *Flickr* [1, 10] as well as

Bing text search results. In the next paragraphs we give a short summary of the main functionalities; more details can be found in [11] and [12].

Collaborative search. Using *LearnWeb2.0*, students and teachers can search for images, video clips, PowerPoint presentations and websites. Figure 1 shows the search results of the query "Wolfgang Borchert". The *LearnWeb2.0* community can collect and create a valuable repository of educational resources to be shared with other colleagues and researchers.

Fig. 1 LearnWeb2.0 search result window

Communication and collaboration. After being added to the *LearnWeb2.0* repository, the resource can be commented on, rated and tagged by all users in the same way as in other Web 2.0 platforms. Ratings reflect the relevance of the resource. By exchanging comments, the students and/or teacher can discuss resource quality, reliability and usability for the course and collaboratively describe the materials more precisely within the learning context. Finally, the selected and rated resources can be used for self-directed learning and re-adapted for designing new course materials.

2.2 Teaching Scenario

Figure 2 shows the different activities carried out in the class, in the laboratory or remotely as a group. As uncontrolled interaction not necessarily triggers learning processes, we used a CSCL script to define some interactions to be carried out during the course. Computer-supported cooperative learning (CSCL) gives guidelines how to support students to cooperate as they interact in groups and how to solve problems in dealing both with the technology and the content [5].

First Phase: preparation [I.1]
At the beginning of the project, the teacher presented the short story "An diesem Dienstag" in class and asked the children to develop a preliminary interpretation through a cooperative *jigsaw* process [2].

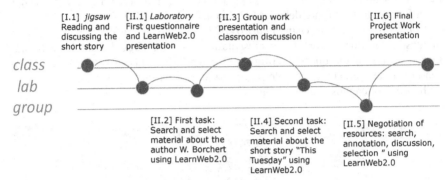

Fig. 2 CSCL script – activities carried out during the project

Second Phase: collaborative project work
Based on these preliminary self-employed working approaches, the task was to use the *LearnWeb2.0* system to collaboratively search and analyse information found on the Web about the author Wolfgang Borchert and his short story "An diesem Dienstag" as an example of postwar literature.

Task

Create a GoogleDoc document on the topic: W. Borchert "An diesem Dienstag" as an example of postwar literature, which will be evaluated as a group-work in class.
Possible items:
- W. Borchert´s biography and other works
- Short story "An diesem Dienstag"
- Postwar literature as an époque, other authors, etc.
- School in National Socialism, everyday life in National Socialism
- Special features / characteristics of short stories

II.1   Laboratory – In order to collect the pupil's preliminary experience regarding the use of the Internet as a tool for learning, we asked the students to fill in an online questionnaire. Then, we presented *LearnWeb2.0* and guided the pupils to register their user accounts and to create small groups of four to six pupils.

II.2   First task – The first application task was to search for suitable material about the author Wolfgang Borchert.

II.3   Group work presentation – In the next lesson, the pupils presented the results of their preliminary selection of resources from the wealth of multimedia resources (e.g. pictures, videos, websites…) available in Internet. Using a resource collected by the pupils as an example, the teacher moderated a discussion about the reliability of online resources.

II.4 and 5   Second task: The groups received a second task to create a presentation (using collaborative Google documents provided in *LearnWeb2.0*) about

Wolfgang Borchert's "An diesem Dienstag" as an example of post-war literature and to present the results to the plenary. Each group worked during two sessions in class and also at home to gather resources, evaluate and comment on them, and to prepare the final presentation.

II.6 Final presentation - Individual groups presented their findings.

We also asked the pupils to fill in a final online questionnaire to give feedback about the use of *LearnWeb2.0* in the project.

3 Evaluation Results

In this section we present some of the feedbacks we received from the pupils in the questionnaires as well as the results of their group work. We focus in particular on the material pupils searched for, selected and collected in their *LearnWeb2.0* groups. Finally we discuss competencies and differences between the groups.

3.1 Questionnaires

With the first questionnaire we collected the children previous experiences in using the Internet.

4 - Do you think that contents you find in the Internet are suitable for learning?
The majority of pupils thinks that Internet content is suitable for learning (82% grade 8, 95% grade 12). Only 5 pupils over the total of 52 don't trust the information in the Internet for learning, or they prefer to read and study on paper. For one pupil the Internet is distracting.

9 - How do you decide, whether to trust the author / source of an article / video in the Internet?
Pupils in grade 8 evaluate sources in a more spontaneous way, while pupils in grade 12 are more critical about the reliability of sources.

Grade 8	Grade 12
-According to my feeling (it has never let me down!)	-I compare several sources, and if the statements are similar, I can work with them.

10 - What do you do when you cannot judge the trustworthiness? (e.g. search for information about the author, ask a friend about his opinion...)?
Pupils in grade 12 seem to be very independent and prefer to search further instead of asking other people. Younger pupils pay less attention or ask their parents or friends.

In the second questionnaire we asked for the pupils´ feedback on the critical use of resources from the Internet and the relevance for learning.

14 - Have you improved your critical use of sources and materials from the Internet during the project?
In general pupils do not feel much improvement on this topic. Possibly, not enough time and emphasis was put on this issue in the project.

Grade 12: „That one should be careful with documents on the Internet was clear to me even before. The way how to treat a source was helpful."

19 - Today the Internet is increasingly used to exchange information. Do you think this is a suitable medium to learn with in school (instead of a teacher or from books)?
In both classes pupils consider Internet as a possible source of information, but they rely on the experience of the teacher to better explain the learning contents and guide them.

Grade 8	Grade 12
-Yes and no: yes, because there are more options to select from, no because the teacher can explain better.	-Only partially. The Internet is faster and offers more information about a topic, however, the trustworthiness and the quality of some sources is questionable and some information are represented poorly or not suitable for school.

3.2 Collection Analysis

The pupils worked in 11 groups, collecting about 10 to 12 resources each, with a minimum of 1 to a maximum of 18. In general, the 8th grade groups collected more resources, and focused more on the author Borchert (e.g. biographies, videos, photos), while the 12th grade pupils collected less resources and focused more on the postwar period and its literature.

In the following paragraphs, we discuss a few examples of representative collections: two from grade 8 and two from grade 12.

Grade 8
Group 1 collected three biographies of W. Borchert: a biography from an introductory course about German literature at the University of Tromso in Norway[1], the German Wikipedia page about Borchert, and a short biography from a project page about German history[2]. In addition, they collected a graph interpreting an essay of Borchert[3]. Further, they collected a YouTube video, as well as seven Flickr photos.

---

[1] http://www.hum.uit.no/ger/kafka/index2.htm
[2] http://www.hdg.de/lemo/einfuehrung.html
[3] http://home.bn-ulm.de/%7Eulschrey/home-unt.html

In their presentation, the pupils included three (!) biographies, a summary and a discussion about "An diesem Dienstag", short summaries about two other essays and two pictures. They also quoted some resources that are not included in their *LearnWeb2.0* group collection, for example from polunbi.de[4], which is contained in another group.

Group 4 relied more on YouTube. They included three videos about the essays "Draussen vor der Tür", "Nachts schliefen die Ratten doch", and "Mein bleicher Bruder", as well as a five-part documentary about Borchert´s life and work. They selected the same Tromso biography as Group 1, another short biography from a site about forbidden literature "Datenbank Schrift und Bild 1900 – 1960", a PDF-summary about "An diesem Dienstag"[5], and a page from the „European Students Review" journal[6] about "Die letzte Zigarette" with citations from Borchert´s short story. They also selected a German Wikipedia article about postwar literature, a page about the history of a school "Die Schule zur NS-Zeit", and a Flickr photo. Finally, they selected a funny video about students making a presentation about postwar literature. The collection is broad and covers a lot of different aspects; it the most extensive collection from grade 8.

The presentation starts with a biography of Borchert (from the LEMO project, selected by Group 1), it continues with a longer summary of "An diesem Dienstag" from Wikipedia, another short biography, a list of some essays, and a summary about the essay "Nachts schliefen die Ratten doch" from Wikipedia. Even though the presentation is fine, the variety of materials they selected in the *LearnWeb2.0* collection is not reflected.

Grade 12

Group 1 is probably the best presentation from the groups of grade 12. They collected two YouTube videos (i.e. „Nachts schlafen die Ratten doch", and „Das Gewitter"), as well as the YouTube documentary about Borchert in five parts. They selected the German and the English Wikipedia article about Borchert, and his biography from polunbi.de; two Wikipedia articles about "Kahlschlagliteratur" and „Nachkriegsliteratur", and one article[7] on the same topic. The resources include another page about "Schulen im Nationalsozialismus" and about "life in national socialism"[8], a website[9] including several citations and texts from Borchert, a page about "An diesem Dienstag"[10], a very extensive website with materials for teaching in schools, the PDF summary about Borchert[11] discussed in class, and a Flickr photo. This is a very good and rich collection of resources.

---

[4] http://www.polunbi.de/
[5] http://www.fundus.org/
[6] http://work-out.org
[7] http://www.literaturwelt.de
[8] http://www.annefrankguide.com
[9] http://www.gratis-gedicht.de
[10] http://www.zum.de
[11] http://www.fundus.org

Indeed, group 1 presentation shows the variety of resources, even though the YouTube material is not cited (probably they have not been really used, due to time constraints).

Group 3 collected fewer resources: one page about postwar literature[12], a You-Tube video about the essay "Todesfuge" from Paul Celan, with no relationship to the topic, one Flickr photo, and the page about the "Inventur2008" school project from the Kantschule Falkensee, not related at all to the topic.

In contrast with the small number of resources, group 3 presentation is good and well structured as the one from Group 1, even though a bit shorter. Judging from the sources they cited in the final presentation, group 3 members did not use much *LearnWeb2.0* for collecting their resources, but searched more often with other web tools, including Google.

4 Conclusions

In this paper we described the project group-work carried out in two German classes during the Winter semester 2011. We used the *LearnWeb2.0* system to explicitly include web resources for teaching German literature at school.

In general we realized that younger pupils were more reactive and active during the experiment; they were curious and willing to experiment with the tool. The older pupils collaborated more seriously; they focused more on the collaborative tasks, but they added a smaller number of resources in their LearnWeb2.0 groups. In total, the children found and selected good quality materials and they created very good presentations, also taking into account the short period of time (eight lessons).

We got useful feedback from both classes and suggestions on how to improve the *LearnWeb2.0* system. In particular children asked for a larger number of search results and more diversification. They appreciated the opportunity of using a personalized environment to carry out their project-work at school, as well as the collaborative writing with GoogleDoc.

Acknowledgements. The authors' work was partially supported by the STELLAR and the TERENCE projects, funded by the European Commission through the FP7 programme. We are grateful as well to the principal of the Leibnizschule in Hanover Mr Veith, and to the students who supported our research on technology enhanced learning.

References

[1] Abel, F., Marenzi, I., Nejdl, W., Zerr, S.: Sharing Distributed Resources in Learn-Web2.0. In: Cress, U., Dimitrova, V., Specht, M. (eds.) EC-TEL 2009. LNCS, vol. 5794, pp. 154–159. Springer, Heidelberg (2009)

[2] Aronson, E., Blaney, N., Sikes, J., Stephan, C., Snapp, M.: The Jigsaw Classroom. Sage Publication, Beverly Hills (1978)

---

[12] http://deutscheliteratur1.blogspot.com

[3] Bellmann, W., Borchert, W.: An diesem Dienstag. In: Bellmann, W. (ed.) Interpreta-tionen. Klassische deutsche Kurzgeschichten, Stuttgart, pp. 39–45 (2004)

[4] Borchert, W.: An diesem Dienstag. In: Bellmann, W.(ed.) Klassische deutsche Kurz-geschichten, Stuttgart, pp. 25–29 (1947)

[5] Dillenbourg, P.: Over-scripting CSCL: The risks of blending collaborative learning with instructional design. In: Kirschner, P.A. (ed.) Three Worlds of CSCL. Can We Support CSCL, pp. 61–91. Open Universiteit., Heerlen (2002)

[6] Heckt, D.H.: Kooperatives Lernen. In: Heckt, D.H., Neumann, K. (eds.) Deutschun-terricht von A bis Z, Braunschweig, pp. 162–165 (2001)

[7] Kellner, D.: Technological transformation, multiple literacies, and the re-visioning of education. E-Learning 1(1), 937 (2004)

[8] Kress, G.: Literacy in the New Media Age, p. 196. Routledge, London (2003)

[9] Lecke, B.: Medienpädagogik, Literaturdidaktik und Deutschunterricht. In: van den Boogaart, M. K. (ed.) Deutschdidaktik. Leitfaden für die Sekundarstufe I und II, Ber-lin, pp. 46–57 (2008)

[10] Marenzi, I., Abel, F., Zerr, S., Nejdl, W.: Social Sharing in LearnWeb2.0. (IJCEELL) 19(4/5/6), 276–290 (2009)

[11] Marenzi, I., Kupetz, R., Nejdl, W., Zerr, S.: Supporting Active Learning in CLIL through Collaborative Search. In: Luo, X., Spaniol, M., Wang, L., Li, Q., Nejdl, W., Zhang, W. (eds.) ICWL 2010. LNCS, vol. 6483, pp. 200–209. Springer, Heidelberg (2010)

[12] Marenzi, I., Nejdl, W.: I Search Therefore I Learn- Supporting Active and Collabora-tive Learning in Language Teaching. In: Okada, A., Connolly, T., Scott, P. (eds.) Col-laborative Learning 2.0: Open Educational Resources, KMI, The Open University UK. IGI Global (in press, 2012)

Towards a Sociosemiotic WebAnalytics: Higher Education TEL Tools Handling Access and Information Extraction in Textually Complex Websites

Cristina Arizzi, Ivana Marenzi, and María Moreno Jaén

Abstract. This paper reports the first stages in the authors' plans to build *WebAnalytics* a sociosemiotically-oriented websearch-and-webshare platform for Higher Education students. It incorporates a model of textuality based on mini-genre theory in which linguistic structures are related to text structures in explicit ways. End-user perceptions of textual and linguistic tie-ups are explored in relation to the functions of search engines and annotations. As space precludes detailed reporting of our field investigations, the experimental tools used are described in relation to the theoretical foundations and the potential for improving their interoperability. In so doing, the paper puts forward an approach to software development that posits a step-by-step integration of existing tools as a way of enacting the planned platform.

1 Introduction

With its very name, *WebAnalytics* signals a concern with a user-oriented perspective of web pages and in particular with the needs of skilled readers (such as Humanities and Social Sciences undergraduates with whom most of our road testing is being undertaken) capable of analysing a web page in terms of its formal composition and using this advantageously when interpreting its content. Such users typically activate reading strategies based on periodicity (e.g. skimming and

Università degli Studi di Messina, Italy
e-mail: `cristina.arizzi@unime.it`

L3S Research Center, University of Hanover, Germany
e-mail: `marenzi@L3S.de`

Universidad de Granada, Spain
e-mail: `mmjaen@ugr.es`

P. Vittorini et al. (Eds.): International Workshop on Evidence-Based TEL, AISC 152, pp. 67–74.
springerlink.com   © Springer-Verlag Berlin Heidelberg 2012

scanning) which contextualize their reading of passages of written texts in relation
to the textual units in which they occur. In this respect, few text linguists would
agree with the assumption that texts can be defined *solely* in linguistic terms i.e. in
relation to the composition and distribution of words, phrases and sentences. The
addition of another perspective, the *textuality* aspect, highlights the need to em-
brace features of websites that would otherwise be discarded. We follow Halliday
and Hasan [10, 11], and many others, in holding that all texts are primarily meaning-making units made up of semiotic ties. We also follow those [2,3,4,5,6] who
hold that web pages, are made up of *web units,* some native to Internet, such as
search engines, headers, menus, bottom bars, logos, headings, text boxes, others
such as running text, titles, subtitles, bylines, headlines and captions deriving from
precursor technologies. They *all* use visual and spatial forms/resources – not just
linguistic ones – tied together cohesively as functional units. Indeed, their welding
of the visual, spatial and linguistic is so strong and distinctive as to make them in-
stantly recognizable across many websites with the result that many (but not all)
web units have the status of *mini-genres* – recurrent objects whose combinations
typically guide website users in their recognition of a particular website's identity,
content and social functions. Second, web units [2,3,4,5,6] are arranged hierarchically with respect to each other on web pages, with some holding a higher status
(genres) and others (mini-genres) standing in a subordinate position in terms of
their contribution to meaning-making. Web units also exist below the mini-genre
level. For example, words and/or arrows help identify the presence of many web
mini-genres (e.g. in Search Engines) on a web page often as a result of their distinctive use of colour, fonts and shapes. Whatever their status, web units co-
contextualize each other in meaning-multiplying ways. Below we present tools for
web users' interactions with websites based on the principles of textuality
and the recognition of web units that lie behind them, that respond to some
shortcomings that we identified in search-and-share functions of the website genre
we chose to analyse: film archives. For reasons of space, we illustrate our
thinking in relation to a one such website, the *Living Room Candidate* (*LRC*)
(*www.livingroom candidate.org*).

2   External vs. Internal Search Engines and Indexing

A previous publication by one of the authors [1] explored the constantly-evolving
LRC site in terms of its value for Political Science students. As a searchable archive for US TV campaign commercials for Presidential Elections from 1952 on-
wards it *ought* to be a paragon of virtue given the co-contextualising functioning
of its web units: the Index Menu which describes the categories the site contains
ought to support the Search Engine's functioning. Indeed, as well as election year
and political party, the main current categorizations on which access is based, *LRC*
includes a helpful category called "Backfire" relating to films using opponents'
words, slogans or images to attack them (e.g. broken promises, flip-flopping on
important matters). Yet of the 300+ plus commercials in the database, *LRC*'s internal Search Engine only identified 9 "Backfire" commercials – hardly the most
convincing application of the site's indexation of categories and not in keeping

with our users' cultural expectations and intellectual needs. Nor did the search word 'discrimination', clearly a keyword in US politics, produce *any* hits. Yet the information was, as Figure 1 shows, there all the time and easily detected using a site-external Search Engine, *MWS* [4]), which supports complex on-the-fly searches, grounded on associative word entries (search words in Figure 1: 'discrimination' *plus* 'Woman/women', 'immigrant/immigrants', 'religious'). Note that we were *not* concerned with identifying the films as such, nor with finding all the cases of the word 'discrimination' but *only* those occurring in the context of other social categories (e.g. minority groups) and as parts of the *descriptions* of the films contained in the archive. Our students were after all concerned both with evolution in attitudes over time and with their efficient retrieval: for them the appearance of words or expressions in a headline is usually more important than their appearance in a footnote. In this respect, thanks to the basic insight that websites implement a hierarchical model of textuality, the *MWS* system also analyses hits (not shown for reasons of space) in terms of the mini-genres in which the detected wordings are found, thus making it possible to provide prioritizations of search results based on the mini-genre ranking.

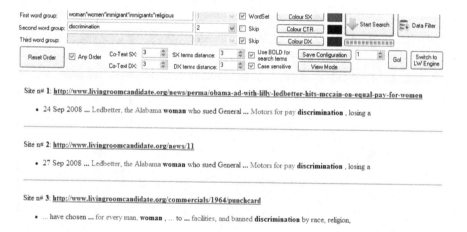

Site n# 1: http://www.livingroomcandidate.org/news/perma/obama-ad-with-lilly-ledbetter-hits-mccain-on-equal-pay-for-women

- 24 Sep 2008 ... Ledbetter, the Alabama **woman** who sued General ... Motors for pay **discrimination** , losing a

Site n# 2: http://www.livingroomcandidate.org/news/11

- 27 Sep 2008 ... Ledbetter, the Alabama **woman** who sued General ... Motors for pay **discrimination** , losing a

Site n# 3: http://www.livingroomcandidate.org/commercials/1964/punchcard

- ... have chosen ... for every man, **woman** , ... to ... facilities, and banned **discrimination** by race, religion,

Fig. 1 Multiple hits with *MWS*

Using site-external search engines to bypass the limitations of site-internal search engines [2, 3, 4, 6] is just a first step. A further step is to build 'external' indexes. As mentioned above, as a website expands, it is very unlikely that a *single* webmaster will be able to maintain the required level of indexing for a single site. An approach which allows the creation of a site-external index applicable to specific genres (e.g. historical film archives) represents a first step in solving this problem. The creation and automatic application of lexical sets involving individual words but also word groups (including collocations and colligations [12]), is nothing new but when associated with web unit hierarchies, represents a way of making category-based searching less improvised and provides instead a mechanism through which index-cum-search categories can be built incrementally over time using machine-learning strategies.

As an example, Figure 2 uses the *MWA* program [4] to show the organization of individual pages in *LRC* in terms of its mini-genre composition. *MWA* automatically recognizes the mini-genres making up web pages (e.g. logos, search engines, headings and so on) [5]. It also allows users to perform queries starting from a word, a collocation or a particular lexical set (Figure 2), and then (currently to a limited experimental degree) automatically filters the search according to the different mini-genres in a web page establishing probabilities and hierarchies in such a way as to classify, describe and above all rank mini-genres in terms of their priority within the website.

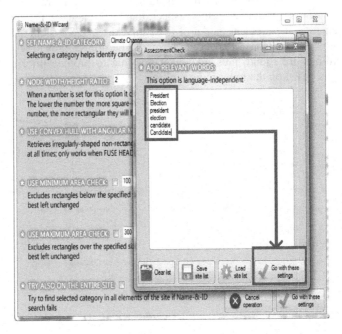

Fig. 2 Linking macro/micro textual organization

With such an approach, HE (Higher Education) users will be able to identify and explore the thematic associations existing between different texts, going beyond the typically chronological (e.g. year-by-year) and A-Z categorisations used in online archives and instead enabling them to analyse such websites in terms of social issues in an associative way, e.g. one in which 'discrimination' is associated with 'women', 'children', 'ethnic minorities', etc.– issues and correlations typically addressed in these sites, but which, as illustrated above, are neither indexed, nor detectable within the site itself.

These automatic procedures will be improved on in the *WebAnalytics* platform; but what has already been achieved highlights the need/capacity to identify social issues typically not mentioned in websites' official indexing but which, as we have shown, can be detected indirectly through the association of automatic macro and micro text search techniques. Using such a technique means that from a user's point of view, instead of a *Google* search, which treats texts as a flat earth (i.e. not hierarchically organized from the standpoint of web unit prioritisation) and retrieves 117 million hits for 'discrimination' and 22 million hits for 'racial discrimination' – useless for many HE end-users; for this word group, or for a word group like 'climate change', our model will give an acceptable number of hits distributed across a web page from Headers (e.g. in Logos) to Footers (e.g. in the *Contact Us* mini-genre). In other words, the reader is able to develop hunches about which hits are really likely to be appropriate for him/her. Such an approach also maintains the associative mechanism we are looking for. In other words, as compared with an internal site engine, or a general external search engine like Google, we are building into our searches an associative and scalar structure corresponding to the nature of texts.

3 The Metatextual Level

Rather uniquely in the TEL (Technology Enhanced Learning) field *LearnWeb2.0*, an integrated environment for searching and sharing Web 2.0 resources, supports HE sharing and managing learning resources distributed across different online platforms [7,8]. Halliday's sociosemiotic theory of the description of language in use – as exemplified by the written, spoken and visual/verbal texts of the Internet – assumes the existence in every text of a constant interplay between different types of meaning: specifically, the experiential, the interpersonal and the textual [6, 7, 8, 9, 10, 11]. Like the *McaWeb* tools (MWS and MWA mentioned above), *LearnWeb2.0* supports exploration of the first two types of meaning thanks to its foregrounding of HE-suitable associations and recontextualisations; as Figure 3 illustrates, query results/resources can be uploaded to a specific community (called a "Group"). The system is capable of identifying resources responding to the search string "Obama campaign US 2000 elections" (Figure 3a) which

Fig. 3 a) Query results; b) Video resource in the LRC Group and annotation fields

represents experiences of the 2000 US elections (experiential meaning) as well as implicit judgments of them (interpersonal meaning). Students signed up to such a group can share their experiences of the US 2000 presidential campaign by collectively building a corpus of texts on this thematic incorporating interpretative and opinion-oriented metatextual annotations.

In the *LRC* group in *LearnWeb2.0* students collected a selection of videos (a *mini-corpus*) in relation to a teacher-assigned task. The work of analytical comparison, so important in HE studies, is further strengthened by the fact that the *Group function* allows members to further *shape* their mini-corpus, and enhance its usefulness, thanks to rating, commenting and other forms of appraisal.

Figure 3b shows how annotations (ratings, tags and comments) can be added in *LearnWeb2.0*. In the case in point, students used comments to discuss the relevance and ranking the video resource in their mini-corpus. In such activities, users build up a vast network of metatextual ties for which a disciplined organization (superior to that found for example in comments sections of online news articles) is required. *LearnWeb2.0* thus provides a metatextual level capable of managing the exponential explosion of commentaries, ratings and so on – meta-texts talking about other texts.

4 Discussion and Conclusions

With what should the design of HE TEL systems start? Textual competence? Linguistic competence? Or both combined? In this paper we pursue the last option. Everybody has a knowledge of textuality and textual composition as defined above and to some extent this is reflected in commercial browsers-cum-search engines (e.g. *Google*'s distinction between such macro-genres as Images, News, Video, Mail, Documents etc). But it could be better exploited in software destined to university graduates given their advanced linguistic *and* textual competences that crucially multiply reading efficiency through their integration.

For this reason, this paper refers to language-based searches in relation to text theory. There are many ways in which the information that websites give can be structured with many new possibilities afforded by Web2.0. We are *not* concerned here with influencing the way in which websites are designed, nor even the fact for example new ways of carrying out political campaigning are being made that transcend the film genres used previously. What we *are* concerned with is developing software procedures that get round the shortcomings and limitation of *current* websites. Without specific workarounds much valuable information is simply inaccessible. Such workarounds need to work at the *text* level as well as the *word* level. For many linguists the term *text* in relation to the Internet is synonymous with the notion of *web units* (many of them *mini-genres*) [3, 5, 6], i.e. units of meaning (e.g. logos, search engines, captions, tables, charts etc.) made up of visual and spatial elements as well as linguistic ones. Moreover, as the above discussion on *mini-genres* demonstrates, texts are not strings of sentences but rather the multiple ties between experiences and judgments about them created by linguistic, visual and spatial resources. The ties involve linkages between texts and meta-texts (e.g. opinions about them). HE students, working in groups, build up comments – in other words their metatextual interpretations – by reconstructing the intertextual ties between texts (films, documents, photos), which exist only as a result of the deductive and associative reasoning described above. While this approach may make things more complex by introducing a text-and-word solution rather than a word-based solution, in another sense it *simplifies* things as in doing so it provides levels of retrieval and commentary consistent with HE needs and activities. The illustrations given above are just one many of these scenarios so far explored which indicate the need to prioritize criteria capable of improving the quality and speed of access to information if TEL is to deliver HE students' requirements vis-à-vis such processes as writing, thinking, reorganizing, critical comparison, work sharing, debate and discussion of different critical perspectives. Commercial software tools often cater for communities whose users run in tens of millions. They often succeed in the process. HE tools need instead to be much more specific and understand the needs of the much more specialised communities they are targeting. Implicit in this short presentation is the belief that HE students need to access the web, appropriating and reshaping the information contained in specific sites in terms of their own multimodal mini-corpora [2], using tools which, through dynamic, real-time on-the-fly procedures, *bypass* websites' internal search engines and produce searches results yet extract information in terms of abstract concepts (e.g. 'discrimination', 'predictions' and textual forms (genres, mini-genres) [2, 3, 4].

Classroom experimentation confirms the appropriateness of a combined text+word-based approach for the proposed *WebAnalytics* TEL platform; it also suggests the need for further fieldwork based on intermediate steps consolidating interoperability between current tools. The goal remains the same: transcending the disproportion between, for example, mighty archives dripping with knowledge and the inadequate site-specific search engines that fail to retrieve this knowledge in HE user-friendly formats and in many cases do not retrieve the information at all.

Acknowledgements. Details of the experimental tools used in this paper can be found on the respective websites Adelex CAT, *http://wdb.ugr.es/~adelex/*. *LearnWeb2.0* (*http://learnweb.l3s.uni-hannover.de/lw/*) and McaWeb tools *http://mcaweb.unipv.it* (EU Grant agreement FP7/2007-2013 n. 231112).

References

[1] Arizzi, C.: Living Room Candidate: A multimodal and cultural analysis of a web archive. In: Baldry, A., Montagna, E. (eds.) Interdisciplinary Perspectives on Multimodality. Campobasso, Palladino (in press)

[2] Baldry, A.: Characterising Transitions in Identity in the Web: Multimodal Approaches and Methods. In: Vasta, N., Riem Natale, A., Bortoluzzi, M., Saidero, D. (eds.) Identities in Transition in the English-Speaking World, pp. 17–38. Forum, Udine (2011a)

[3] Baldry, A.: Multimodal Web Genres: Exploring Scientific English. IBIS, Como (2011b)

[4] Baldry, A., Gaggia, A., Porta, M.: Multimodal Web Concordancing and Annotation. An Overview of the MCAWEB System. In: Vasta, N., Natale, A.R., Bortoluzzi, M., Saidero, D. (eds.) Identities in Transition in the English-Speaking World, pp. 39–60. Forum, Udine (2011)

[5] Baldry, A., Thibault, P.J.: Multimodal Transcription and Text Analysis: A Multimodal Toolkit and Coursebook with Associated Online Course. Equinox Publishing, London (2006)

[6] Cambria, M.: Websearching and corpus construction of online news sites in ESP:Government leaders on show at G8 Summits. ESP Across Cultures 8 (in press)

[7] Marenzi, I., Zerr, S., Abel, F., Nejdl, W.: Social Sharing in LearnWeb2.0. International Journal of Continuing Engineering Education and Life-Long Learning, IJCEELL 14(4, 5, 6) (2009)

[8] Marenzi, I., Kupetz, R., Nejdl, W., Zerr, S.: Supporting Active Learning in CLIL through Collaborative Search. In: Luo, X., Spaniol, M., Wang, L., Li, Q., Nejdl, W., Zhang, W. (eds.) ICWL 2010. LNCS, vol. 6483, pp. 200–209. Springer, Heidelberg (2010)

[9] Halliday, M.A.K.: An Introduction to Functional Grammar, 2nd edn. Arnold, London (1985,1994)

[10] Halliday, M.A.K., Hasan, R.: Cohesion in English. Longman, London (1976)

[11] Halliday, M.A.K., Hasan, R.: Language, Context and Text: a social semiotic perspective. OUP, Oxford (1989)

[12] Moreno Jáen, M.: A corpus-driven design of a test for assessing the ESL collocational competence of university students. International Journal of English Studies 7(2), 127–147 (2007)

Adapting with Evidence: The Adaptive Model and the Stimulation Plan of TERENCE

Mohammad Alrifai, Rosella Gennari, and Pierpaolo Vittorini

Abstract. TERENCE is an FP7 ICT European project that aims at developing an adaptive learning system for supporting learners and educators. The TERENCE learners are 7-8 to 11 year old children with poor reading comprehension skills. The TERENCE educators are primary-school teachers, support teachers and parents. This paper describes the stimulation plan for the TERENCE learners, based on clinical practice, and the adaptive learning model of TERENCE that stems from the stimulation plan. In other words, the design of the model follows the evidence based design.

1 Introduction

A *learning management system* (LMS) is a suite of functionalities designed to deliver, track, report on and manage learning content, learners' progress and learners' interactions. LMSs can apply to very simple course management systems, or highly complex enterprise-wide, distributed environments. Within LMSs, the ability to tailor instruction to individual and organisational needs become a crucial issue, thus leading to the development of the so-called *adaptive learning systems* (ALSs) [4]. Differently from the other LMSs, an ALS features the *adaptation engine* that actually personalises the learning process, based on the *adaptation model*.

TERENCE is a project that aims at designing and developing the first *adaptive learning system* (ALS) for poor comprehenders, i.e., primary school children that

Mohammad Alrifai
L3S Research Center, Leibniz University of Hanover, Appelstr. 9A, 30167 Hanover, Germany

Rosella Gennari
Free University of Bozen-Bolzano, Faculty of Computer Science, P.zza Domenicani 3, 39100 Bolzano, Italy

Pierpaolo Vittorini
University of L'Aquila, Dep. of Internal Medicine and Public Health
V.le S. Salvatore, Edificio Delta 6 – 67100, L'Aquila, Italy

P. Vittorini et al. (Eds.): International Workshop on Evidence-Based TEL, AISC 152, pp. 75–82.
springerlink.com   © Springer-Verlag Berlin Heidelberg 2012

have well developed low-level cognitive skills (e.g., word decoding), but that have problems with deep text comprehension [5, 7]. These are the TERENCE *learners*.

This paper focuses on the adaptation model of the TERENCE system for the TERENCE learners. The model has rules that formalise the stimulation plan for the learners, which is the other main contribution of this paper. The plan is based on clinical practice, and has been informally specified through brainstorming meetings with therapists expert of the TERENCE learners. In other words, the design of the adaptive model of TERENCE follows the *evidence based design* (EBD). This paper ends with a recap conclusive section.

2 Background

The conceptual model of the TERENCE ALS is modularised into (1) the user models, including the learner sub-model specifying and structuring the requirements of poor comprehenders, (2) a domain model that structures the learning material, (3) the adaptation model for the adaptation learning process, specifying the rules correlating concepts of the domain model and the user model. The user and domain models are only briefly sketched here, and more extensively described in [2].

The User Model. The user model has, as submodels, the model for the expert (e.g., psychologists), that for the educators and, most importantly, that for the learner, which structures the data concerning the TERENCE learners. The learner model also structures the so-called *reading comprehension* (RC) skills of a learner. Each RC skill has also different RC levels.

The Domain Model. The reading material of the TERENCE ALS is given by stories, in English and in Italian, adapted to the specific requirements of poor comprehenders. Stories are grouped into thematic coherent units, that is, books. The reading interventions of the system are interactive smart games centred around inference-making skills that foster the development of deep text comprehension of poor comprehenders; the interventions are rooted in the literature of psychologists and therapists working with poor comprehenders, as well as educators. See [3]. The books, stories and games are structured as in the domain model of the system. In particular, in the domain model of TERENCE, each story is associated to the RC skills, specified by the learner model, at a certain RC level. Moreover, each story is associated to a set of smart games. Each smart game serves to evaluate a specific RC skill.

The Adaptation Model. The adaptation model of an ALS is a set of rules that describes how knowledge stored and structured as in the user and domain models can be used for providing adaptive learning experience to the learners. In the remainder of this paper, we describe the adaptation model of the TERENCE ALS for the TERENCE learners. Firstly, we describe the evidence for the model, namely, the stimulation plan for the TERENCE learners. Then we move on to explain specific adaptation activities, based on the stimulation plan: the book and story recommendation; the book adaptation activities; the learner profile's update.

3   Stimulation Plan for Adaptation

The stimulation plan is mainly based on [6]. In the following, we give the essential ingredients of the stimulation plan, and that are turned into concepts of the adaptation model in the subsequent section.

3.1   Cycles and Sessions

A *session* of the stimulation plan consists in reading a story of a book, and then resolving the correlated games. Reading a story is a *reading activity*. Playing a game is a *playing activity*. In general, in a session, games are ordered as follows. First a subset of smart games is proposed. Then the learner plays with relaxing games, which are unrelated to the story and have a relaxing and distracting effect. Then, another subset of smart games is proposed.

Each week of the plan should foresee at least two or three sessions: the higher the number of sessions is in a week, the higher the number is of read stories and played games, the stronger the stimulation is.

A *cycle* consists of 2–3 sessions per week, lasting 2–3 months, with a brief suspension of c.a one week. The longer the cycles, the shorter the suspension, the stronger the stimulation. Evidence-based clinical practice suggests a suspension of circa 2 weeks. See also Fig. 1 for an example of a cycle of 2 months and 2 weeks, with 10 sessions.

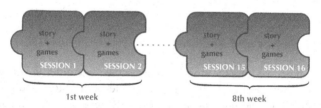

Fig. 1 A cycle of c.a 2 months and 2 weeks, divided into 10 sessions, each lasting one week

3.2   Measures of Activities

Reading and playing activities have different types of measures, described as follows.

The reading activity has diverse logged times. The *reading time* is the time spent by a learner in reading a story of a session, whereas the *maximum reading time* of a story is the maximum time allowed for reading the story, independently of the learner. The *average reading time* for a learner is the average of the reading times the learner has spent in reading stories across the already run sessions.

For the playing activity in a session, we should at least take care of two different types of measures: concerning time; concerning resolutions of games. As for time, when playing in a session, we have the following measures:

– the *resolution time* of a game in the session is the time spent by the learner for resolving the game;
– the *maximum resolution time* of a game in the session is the maximum time allowed for resolving the game, independently of the learner;
– the *average resolution time* for a learner in the session is the average of the resolution times the learner has so far taken for resolving the played games of the session.

All are updated after playing with a game. As for the resolutions of a learner, while playing in a session, we have the following measures for the session's games that evaluate a certain RC skill:

– the *accuracy* ratio is the number of games for that RC skill correctly resolved so far in the session, divided by the total number of games for that RC skill so far played;
– the *omission* ratio is the number of games for that RC skill that are so-far skipped or unresolved within the maximum resolution time in the session, divided by the total number of games for that RC skill so far played.

Given the aforementioned measures, clinical practice suggests that, at the end of a session, we can improve and update the learner's RC level of the RC skill evaluated by the played smart games if

1. the omission ratio decreases, and the accuracy ratio increases,
2. possibly, the average resolution time decreases, and the average reading time decreases.

3.3 Smart Games and Sessions

Smart games should address the story's events in the same order in which these are presented in the story. For each RC skill, the RC level of a story and games in the first session should be the same as the learner's RC level, or even slightly inferior than this, so as not to frustrate the learner. According to the updates to the learner's RC skill at the end of a session, the system can then increase the related RC level of the story and games in the subsequent sessions.

The first session is likely to be slightly different than the subsequent sessions, because in that session the learners need to acquaint with the system. Accordingly, the first session will mainly consist of the training to the system. For instance, the learner will get familiar with the system's interface, and the type of interactions required by the TERENCE games. In particular, the resolution times during the first session and the accuracy ratio are likely to depend on the concurrent training with the system. During the other sessions, the resolution times of the session's games should become independent from this aspect.

If, in a session, the learner makes a significant number of mistakes in resolving the smart games of the story (that is, the learner's accuracy is low or the learner's omission is high), the system:

- shall propose the same story in a simplified version in a subsequent session with the system;
- shall propose easier games, or games with other feedback than the correctness of the resolution.

Another important part of the stimulation consists indeed in the feedback to the resolution of a game, which is adapted to the learner and the session. In the first session, the only feedback is a so-called consistency feedback, textually and visually given: yes, the resolution is correct; no, it is not. In the later sessions, the feedback can change according to the profile of the learner. For instance, if in the previous session the learner had a low profile in a certain type of games, the feedback to the resolution of the same type of games in the current session will be of type explanatory, offering cues for resolving the game in case the learner's time in resolving the game is higher than the maximum resolution time.

4   From the Stimulation Plan to the Adaptation Model

The stimulation plan lends itself naturally to the design of the adaptation model of the TERENCE ALS. In particular, it provides information for the recommendation of books to the learner, for the navigation of the book's stories, for the update of the learner's profile according to the RC skill measured by certain smart games in a session with the system. In this section, we consider each of such issues separately.

4.1   Recommendation of Books

The TERENCE ALS is capable of recommending adequate books for the learner by matching concepts and data related to the book model with concepts and data related to the learner. As for this, relevant concepts are the learner's interest, preferences and demographic characteristics, structured as in the learner model.

4.2   Navigation of a Book

Traditionally, in game design, a linear book has a linear narrative plot: stories of the book are related through a finite linear order. Instead, a branching book has got several branching narrative plots. A branching book can thus be associated to a rooted *directed acyclic graph* (DAG). The root of the DAG is labelled by the starting story of the book, and the ending stories label the leaves. Each node in the DAG is a story, i.e., a small self-contained part of the whole book. An edge from story x to story y in this rooted DAG means that reading story x is required to have an access to story y. See [1].

The logical design of the TERENCE books follows a branching book design. However, in a TERENCE branching book, each story is available in different versions. Each version is associated to a different RC level. The root of the DAG associated to a TERENCE book is labelled with the version of the starting story having

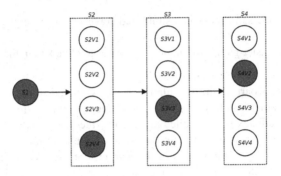

Fig. 2 The branching structure of a TERENCE book

the RC level expected for the learner. The DAG is then structured in layers: the nodes of the n-th layer are each labelled by a different version of the same story, starting from that at RC level 1. We have only the following constraint on the edges of the DAG of the TERENCE books: edges are from nodes of layer n to nodes of layer $n + 1$.

The learner, instead, is shown only a branch of the DAG. This means that the learner experiences a linear book, with a starting story and a final one. Moreover, in each session, the learner is shown only the versions of the still unread stories at his or her RC level. At the interface layer, the learners sees a pictorial representation of the branch: the nodes of the branch, that is, the different versions of the available stories of the book at his or her level, are visualised by showing the stories' environments.

Fig. 2 shows a simple case with four stories, and hence four layers in the DAG. The book starts with the first story S_1 at the expected RC level; this labels the root of the DAG of the book. Then, the second story S_2 is made available in four versions, i.e., $S_2V_1, S_2V_2, S_2V_3, S_2V_4$; each of this four versions labels a different node at layer 1. Similarly for the others. From the first story, depending on the profile of the learners, the adaptation consists in providing the learner with the version of the second story that matches the learners' current RC level. The red nodes show a path throw the DAG, and corresponds to the linearisation of the branching book that the learner experiences right after choosing the book.

Such a branching book design allows the learner to have a story adapted to the learner's RC level at every session.

4.3   The Learner Profile's Update

As the learners' RC skills can change during the learning process lifecycle, their profiles must reflect such changes. Therefore, the TERENCE ALS keeps track of the data of the learners' interactions with the system, and updates the learners' profiles accordingly.

The games in TERENCE are designed for assessing an RC level, or for relaxing. Therefore, a key interaction that influences the learner's profile is the game playing activity. In line with the stimulation plan of Sect. 3, when the system observes that, at the end of a session, a learner has demonstrated a progress in resolving the games with an RC difficulty level not inferior to his or her currently stored level, the system updates the learner's level for this particular RC skill accordingly.

Notice that updating the RC skills of a learner automatically triggers an adaptation of the book's stories, that is, which versions of the stories should appear along the path that the learner is shown, as described above in Subsect. 4.2.

Moreover, by solving a game successfully, that is, by increasing his or her accuracy rate in a session, a learner gains points, which in turn leads to updating the score of the learner's avatar, that is, the learner's virtual representation, at the end of the session.

Fig. 3 shows some of the key concepts and attributes that are involved in this adaptation process in a session. Those based on the stimulation plan of Sect. 3 are highlighted in blue. For instance, the points acquired in a session's game trigger updates to the accuracy ratio of the learner. The update to the omission rate of the session is correlated to whether a session's game is skipped, or to the learner's resolution time for the game with respect to the allowed maximum resolution time in the session. In turn, the updates to the accuracy rate and the omission rate can trigger updates to the RC level of the learner at the conclusion of the session.

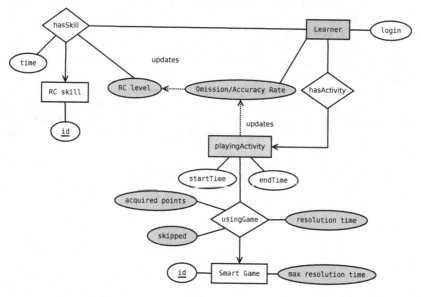

Fig. 3 The learner profile's update in a session

5   Conclusions

This paper described the adaptive model of the TERENCE system, and its EBD. The first phase of the design process, conducted by experts of the TERENCE consortium, focused on the analysis of the context of use and user requirements. In particular, this process allowed us to informally specify the stimulation plan of the TERENCE system for its learners mainly through brainstorming meetings and contextual inquiries with therapists. The plan is rooted in daily clinical practice, based on evidence. The plan was then turned into the specification of the adaptation model, which thus follows the EBD. Both the plan and key ingredients of the model were explained in this paper. Future work will consider the adaptation of the feedback to the learner, as well as the adaptation rules for the other main users of the TERENCE system, namely, experts and educators.

Acknowledgments. The authors' work was supported by TERENCE project, funded by the EC through the FP7 for RTD, Strategic Objective ICT-2009.4.2, ICT, TEL. The contents of the paper reflects only the authors' view and the EC is not liable for it. The second author work was also funded through the CRESCO project, financed by the Province of Bozen-Bolzano.

References

1. Adams, E.: Fundamentals of Game Design. New Riders (2010)
2. Alrifai, M., Gennari, R., Tifrea, O., Vittorini, P.: The Domain and User Models of the TERENCE Adaptive Learning System. In: Proc. of eb-TEL 2012. Springer (2012)
3. Arfé, B., Boscolo, P., Boureux, M., Carretti, B., Di Mascio, T.L.L., Oakhill, J., Tifrea, O.: State of the Art of Methods for the User Analysis and Description of Context of Use. Tech. Rep. D1.1, TERENCE project (2011)
4. Brusilovsky, P., Millán, E.: User Models for Adaptive Hypermedia and Adaptive Educational Systems. In: Brusilovsky, P., Kobsa, A., Nejdl, W. (eds.) Adaptive Web 2007. LNCS, vol. 4321, pp. 3–53. Springer, Heidelberg (2007)
5. Cain, K., Oakhill, J.V., Barnes, M.A., Bryant, P.E.: Comprehension Skill, Inference Making Ability and their Relation to Knowledge. Memory and Cognition 29, 850–859 (2001)
6. Di Giacomo, D., Di Mascio, T., Gennari, R., Melonio, A., Tifrea, O., Vittorini, P.: State of the art and design of novel intelligent feedback. Tech. Rep. D4.1, TERENCE project (2011)
7. Marschark, M., Sapere, P., Convertino, C., Mayer, C.: W.L., Sarchet, T.: Are Deaf Students' Reading Challenges Really About Reading? (in Press)

The User and Domain Models of the TERENCE Adaptive Learning System

Mohammad Alrifai, Rosella Gennari, Oana Tifrea, and Pierpaolo Vittorini

Abstract. TERENCE is an FP7 ICT European project that aims at developing an adaptive learning system for supporting learners and educators: the TERENCE learners are 7-8 to 11 year old children with poor reading comprehension skills; the TERENCE educators are primary-school teachers, support teachers and parents. The analyses of the context of use and requirements of the TERENCE system are based on real data. Therefore also the design of the conceptual model of the TERENCE system is based on real data. This paper describes the domain and user model of the TERENCE system, and its evidence-based design process.

1 Introduction

Developing the capabilities of children to comprehend written texts is key to their development as young adults. From the age of 7-8 until the age of 11, children develop as independent readers. Nowadays, more and more children in that age range turn out to be *poor (text) comprehenders*: they demonstrate difficulties in deep text comprehension, despite well developed low-level cognitive skills like vocabulary knowledge, e.g., see [4] for hearing poor comprehenders, and [13] for deaf poor comprehenders. TERENCE is an FP7 ICT project that aims at designing and developing the first adaptive learning system for poor comprehenders. Its reading material (in English and in Italian languages) are stories adapted to the specific requirements of poor comprehenders, and its reading interventions are interactive games centred

Mohammad Alrifai
L3S Research Center, Leibniz University of Hanover, Appelstr. 9A, 30167 Hanover, Germany

Rosella Gennari · Oana Tifrea
Free University of Bozen-Bolzano, Faculty of Computer Science, P.zza Domenicani 3, 39100 Bolzano, Italy

Pierpaolo Vittorini
University of L'Aquila, Dep. of Internal Medicine and Public Health
V.le S. Salvatore, Edificio Delta 6 – 67100, L'Aquila, Italy

P. Vittorini et al. (Eds.): International Workshop on Evidence-Based TEL, AISC 152, pp. 83–90.
springerlink.com   © Springer-Verlag Berlin Heidelberg 2012

around reasoning skills that foster the development of deep text comprehension, both accompanied by adequate visual aids.

This paper starts by briefly overviewing the state of the art of adaptive learning systems, and the methodology adopted for designing the TERENCE system. With the preliminaries out of the way, the paper zooms in on the conceptual model of the TERENCE system, and then delves into its main sub-models. The paper ends with a recap conclusive section.

2   Related Work on Adaptive Learning System

The ability of a learning system to tailor instruction to individual and organisational needs is a crucial issue, and lead to the development of the *adaptive learning systems* (ALSs) [3]. The conceptual model of an ALS usually includes, as sub-models, the *user model* for the users's data, the *domain model* for the learning materials' data, the *adaptation model* with rules that, given the previous models, provide the actual adaptation.

Of the aforementioned conceptual sub-models, the user model is the one that has got more attention and proof-of-concept implementations. Aside from distribution, scalability and performance aspects, the principal motivation for the development of user models is to characterise an individual user. Traditionally the main user features that are considered in ALSs include user's knowledge, background, interests, goals and preferences. User individual traits, such as personality factors, cognitive factors and learning styles, were given attention by the research community, albeit with relatively few practical uses. The KBS-Hyperbook [11] and TRAILS projects [10] based their modeling on (reasoning over) logged user actions. In the AHA! project [6], on the other hand, user actions are typically not logged but immediately translated into higher-level user model information. Various adaptive systems take into account the user's context [12], such as the user's location, platform and bandwidth [2]. In several ALSs, the user model is an overlay of the given domain ontology, which associates the users' knowledge or interests with a concept in the domain. However, few user models described in the literature provide a general representation of users in the form of an ontology, one of them being the General User Model Ontology (GUMO) [9]. For more details and examples, we refer to a survey carried out by [3].

3   The Design Process

The TERENCE system is being developed by following the user-centered design (UCD), and evidence-based design (EBD). Generally speaking, the UCD places the users at the centre of the design process, which is iteratively repeated through subsequent refinements of the requirements until attaining the usability of the system. The EBD always requires the usage of evidence-based data in such a design process.

In the first year of life, the TERENCE consortium specified the requirements of the system. The specifications are based on the state of the art as well as on the analysis of data collected mainly through: brainstorming meetings with experts;

contextual inquiries with educators and learners. In the former case, the data are purely qualitative. In the latter case, they are both qualitative and quantitative. See also [7]. The first specifications were in a semi-formal tabular format and gave us:

1. the characteristics of the users, like their reading comprehension skills or interests in books;
2. tasks, like successful reading interventions by class teachers for improving reading comprehension;
3. the environments, divided into organisational, physical and socio-cultural characteristics that may influence the usage and acceptance of the system.

Our design of the current version of the conceptual model of the TERENCE ALS is based on such semi-formal specifications, as well as on the state of the art of conceptual modeling for ALSs:

1. its domain model structures the learning material,
2. its user models specifies and structures information concerning its users, in particular, its learners,
3. its adaptation model specifies the rules correlating concepts of the domain model and the user model.

In the following, we specify the main concepts and relations of the user and domain models, which find their evidence in the semi-formal specification reported above. The adaptation model is reported in another paper [1]. The domain and user models are here represented with an Entity-Relationship (ER) diagram [5]. We opted for ERs for several reasons. First of all, ERs are easy to understand, and easy to explain to all the members of the TERENCE heterogeneous consortium. Moreover, the ERs serves as the concept scheme of the underlying databases. As in the UCD the design of interactive systems evolves cyclically, also the domain and user models of the TERENCE system needs to be updated iteratively according to the updates to the requirements, and ERs are easy to manage and update. However, ERs are only the building blocks, so to speak: they are used to create the OWL ontology of the models, as OWL enables the flexibility, interoperability and reusability of the model, besides reasoning services like consistency checking and deduction.

4   The User Model

The user model is made of three main sub-models, one for each end-user of the TERENCE system: the *learner model* for the learners of the TERENCE system, the *educator model* for the educators of the TERENCE system, namely, teachers and parents of the TERENCE learners, the *expert model* for the expert users (e.g., linguists) interacting with the TERENCE system. In the remainder, we mainly focus on the learner model, which captures information about the TERENCE learners that is relevant for the adaptation process. The structure of the TERENCE learner model consists of three main sub-models: the first for general data; the second for domain dependent data; the third for interaction data. They are described as follows.

The first sub-model represents general data about the learners such as their name, contact information, personal characteristics, and interests. The main source for this sub-model is GUMO[9].

The second sub-model of the learner model represents domain dependent data about the learner's *reading comprehension* (RC) skills. The ER for the RC skills analysis whether the problems are at the word level, sublexical level, sentence level, or entire text level. Such data and their representation are important for a competency-based classification of the learners. In particular, each learner can choose among several *avatars*, which work as a virtual representation of the learner and his/her progress. The learner chooses one of the available avatars before starting reading and playing with TERENCE. As the learner proceeds in reading and playing, he/she gains points, which allows him/her to obtain more visual attributes for his/her avatar. Such attributes are represented with the score entity in the ER. By analysing the learner's skills, the system will be able to suggest appropriate content for the learner as well as provide relevant input to the educators. The main source is the semi-formal specification of the learners' characteristics deriving from the aforementioned requirement analysis.

The third and last sub-model of the learner model includes data about the learner's interaction with the system, that is, logs about the learner's activities such as reading stories or playing games are stored in the model. By storing these data the system will be able to help the educators in keeping track of their learners' progress. Figure 1 shows the ER diagram of the learner model.

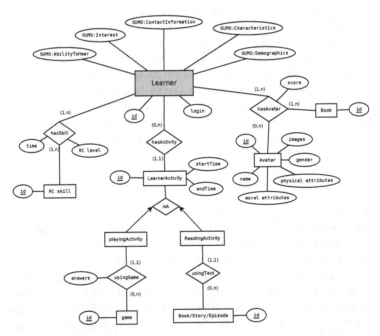

Fig. 1 The current learner model

5   The Domain Model

The TERENCE Domain Model consists of two main sub-models: the reading material model, made of the book and story models; the playing material model, currently consisting of the game model. The main source for the domain model is the semi-formal specification of the learners' tasks, namely, reading and playing, and deriving from the requirement analysis.

5.1   The Reading Material Models

Since the learners will have different backgrounds, interests and reading comprehension skills, the system will provide stories with different characteristics and difficulty levels, which depend on the RC levels specified as in the learner model. Moreover, stories are grouped into coherent narrative units, named books. Figures 3 and 3 show the ER diagrams of the story and book models, making up the reading material model. In the following we describe the main concepts of this model.

Book. The entity Book represents one of the written books that are available in the repository of the system. Each Book has the following main attributes:

- id: a unique id;
- name: the title of the Book;
- age range: the intended age range for reading the Book;
- genre: the genre of the Book's stories (e.g. fantasy, gothic, adventure, etc.);
- abstract: the main topic of the Book;
- visualMap: each Book in TERENCE will have a visual spatial map that illustrates the environments of the Book's stories. By interacting with the visual map, the learner can thus view the stories and select one of them.

Moreover, each Book has an associated set of main characters, those that recur in the Book's stories. Each character has textual and graphical descriptions.

Story. This entity represents a story of the book with main attributes:

- id: a unique id for the story;
- name: the title of the story;
- abstract: the abstract of the story.

In addition, each story, except the last, has a link to the subsequent stories.

RC (reading comprehension) Skill. Each story will be analysed using natural language processing tools such Coh-Metrix [8] (Duran, et al. 2006) in order to assess the story's difficulty level. This is expressed in terms of the RC skills and the minimum level of each such skill in order to comprehend the text, that is, the RC level.

Episode. According to the analysis of tasks mentioned in Section 3, the analysis of stories is usually conducted episode by episode, following their textual order in the story. Therefore we include Episodes as entities related to story in the ER.

Game. Each story is associated with a set of games for stimulating and assessing the reader's comprehension of the story, and described below.

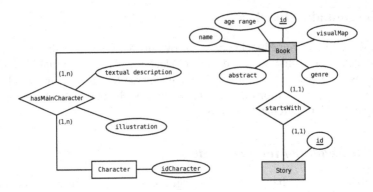

Fig. 2 The current book model

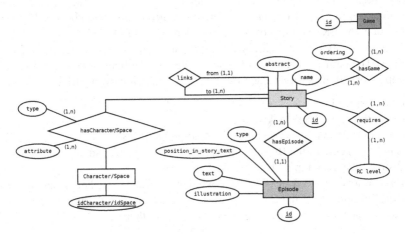

Fig. 3 The current story model

5.2   The Game Model

According to the analysis of tasks and users' characteristics sketched in Section 3, TERENCE needs to have two main types of games: smart games, relaxing games. A smart game in TERENCE consists a cognitively demanding reasoning problem concerning the story's main events, characters and their relations. Relaxing games allow the learner to have breaks from reading, and so maintain the attention alive. The ER diagram of the Game Model is shown in Figure 4, and its main entities are described as follows.

Game. The entity Game represents a game of a story with main attributes:

– id, a unique identifier;
– name, that is, the title of the Game;
– the problem (Task in the ER) that the learner has to deal with;

- the maximum number of Points a player can get for resolving the problem correctly;
- the Maximum Resolution Time, for resolving the problem;
- the Possible and Correct Resolutions of the problem.

Stimulation target. Each Game has a certain stimulation target (stimulationTarget), which is related to one or more RC (reading comprehension) skills. Each Game has also a certain difficulty level (the attribute RC level) for each of the evaluated RC skills. The difficulty level is the RC skill level required by the game.

Smart game and Relaxing game. Each Game can be either a Smart game, which stimulates reading comprehension, or a Relaxing game, which does not.

Story. Each story is associated to Games via the hasGame relation. This has the ordering attribute for setting the right alternation between Smart and Relaxing games.

Feedback. Each game may have diverse types of Feedback, depending on the learner's profile. The Feedback is given only if the Game is wrongly resolved.

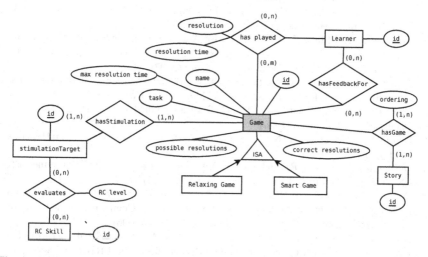

Fig. 4 The current game model

6 Conclusions

The TERENCE system is developed following the UCD and the EBD. The first phase of the project focused on the analysis of context of use and user requirements. This ended into a semi-formal specification of the characteristics of the TERENCE's users and tasks. Our design of the domain and user models of the TERENCE ALS, which is the focus on this paper, is based on such specifications. In particular, our study of the state of the art for user modeling in ALSs and the specification of the characteristics of the users allowed us to define: the learner model; the educator model; the expert model. The analysis of the tasks allowed us to specify the domain model which structures the learning material of TERENCE, and is divided into the book, story and game sub-models. Last but not least, in the UCD, the design of interactive systems evolves cyclically. Therefore also the specification of the conceptual model of TERENCE will be updated iteratively according to the updates to the requirements.

Acknowledgments. The authors' work was supported by TERENCE project, funded by the EC through the FP7 for RTD, Strategic Objective ICT-2009.4.2, ICT, TEL. The contents of the paper reflects only the authors' view and the EC is not liable for it. The second author work was also funded through the CRESCO project, financed by the Province of Bozen-Bolzano.

References

1. Alrifai, M., Gennari, R., Vittorini, P.: Adapting with Evidence: the Adaptive Model and the Stimulation Plan of TERENCE. In: Proc. of eb-TEL 2012. Springer (2012)
2. Brusilovsky, P.: Adaptive Hypermedia: From Intelligent Tutoring Systems to Web-Based Education. In: Gauthier, G., VanLehn, K., Frasson, C. (eds.) ITS 2000. LNCS, vol. 1839, pp. 1–7. Springer, Heidelberg (2000)
3. Brusilovsky, P., Millán, E.: User Models for Adaptive Hypermedia and Adaptive Educational Systems. In: Brusilovsky, P., Kobsa, A., Nejdl, W. (eds.) Adaptive Web 2007. LNCS, vol. 4321, pp. 3–53. Springer, Heidelberg (2007)
4. Cain, K., Oakhill, J.V., Barnes, M.A., Bryant, P.E.: Comprehension Skill, Inference Making Ability and their Relation to Knowledge. Memory and Cognition 29, 850–859 (2001)
5. Chen, P.: The Entity-Relationship Model–Toward a Unified View of Data. ACM Trans. Database Syst. 1(1), 9–36 (1976)
6. De Bra, P., Smits, D., Stash, N.: The Design of AHA! In: ACM Conference on Hypertext and Hypermedia (2006)
7. Di Mascio, T., Gennari, R., Melonio, A., Vittorini, P.: The UCD and EBD in the TERENCE project. In: Proc. of eb-TEL 2012. Springer (2012)
8. Duran, N., McCarthy, P.M., Graesser, A.C., McNamara, D.S.: Using Coh-Metrix Temporal Indices to Predict Psychological Measures of Time. In: Conference of the Cognitive Science Society, pp. 190–195. Cognitive Science Society (2006)
9. Heckmann, D., Schwartz, T., Brandherm, B., Kröner, A.: Decentralized User Modeling with UserML and GUMO. In: Workshop on Decentralized Agent Based and Social Approaches to User Modelling (DASUM 2005), pp. 61–65 (2005)
10. Heller, J., Levene, M., Keenoy, K., Albert, D., Hockemeyer, A.: Cognitive aspects of trails: A stochastic model linking Navigation Behaviour to the Learner's Cognitive Stateavigation Behaviour to the Learner's Cognitive State. In: Trails in Education. Technologies that Support Navigational Learning. Sense Publishers (2007)
11. Henze, N., Nejdl, W.: Adaptivity in the KBS Hyperbook System. In: 2nd Workshop on Adaptive Systems and User Modeling on the WWW, Workshop held in Conjunction with the World Wide Web Conference (WWW8) and the International Conference on User Modeling (1999)
12. Jameson, A.: Modeling both the Context and the User. Personal and Ubiquitous Computing 5, 29–33 (2001)
13. Marschark, M., Sapere, P., Convertino, C., Mayer, C.: W.L., Sarchet, T.: Are Deaf Students' Reading Challenges Really About Reading? (in Press)

Visual Representations of Narratives for Poor Comprehenders

Tania Di Mascio, Rosella Gennari, Alessandra Melonio, and Pierpaolo Vittorini

Abstract. Poor comprehenders are children with specific reading impairments. Several reading interventions for them, evaluated in the literature, use images for representing textual information in narratives. Our paper overviews several such studies, and highlights current findings and shortcomings in the literature. The results of the overview can be taken up for designing specific evidence-based visual representations of narrative information for poor comprehenders.

Keywords: evidence-based design, visual representation, users with special needs.

1 Introduction

Nowadays, more and more children in that age range turn out to be poor (text) comprehenders: they demonstrate difficulties in deep text comprehension, despite well developed low-level cognitive skills like vocabulary knowledge, e.g., see [1, 2]. In particular, they seem to have problems in making inferences for answering questions like who, what and where, as well as for correlating events narrated in the text in a coherent causal-temporal model.

Pictorial aids can help children to comprehend specific textual information, and answer such questions, problematic for poor comprehenders. In this paper we overview several experimental studies which assess the effect of pictorial aids on the text comprehension of poor comprehenders.

The skeleton of this paper is briefly given as follows. The subsequent section lays the groundowrk: it overviews the common functions of images for texts. Next, the paper reports on field studies with poor comprehenders, experimentally

Tania Di Mascio
University of L'Aquila, DIEI, Via G. Gronchi, 18 – 67100, L'Aquila, Italy

Rosella Gennari · Alessandra Melonio
Free University of Bozen-Bolzano, CS Faculty, P.zza Domenicani 3, 39100 Bolzano, Italy

Pierpaolo Vittorini
University of L'Aquila, MISP, V.le S. Salvatore, Edificio Delta 6 – 67100, L'Aquila, Italy

P. Vittorini et al. (Eds.): International Workshop on Evidence-Based TEL, AISC 152, pp. 91–98.
springerlink.com   © Springer-Verlag Berlin Heidelberg 2012

evaluating the role of visual representations with specific functions for text comprehension. The paper concludes by assessing the evidence available in the literature for recommending specific visual representations for key features of events and causal-temporal relations in texts, whose textual comprehension is problematic for poor comprehenders.

2 Visual Representations and Narrative Comprehension

Our compact overview of the literature concerning visual representations for narrative comprehension starts with the work reported in [4], and extended in [5, 6]. Levie et al. analyse the literature on pictures for the comprehension of text, and classify pictures into five types of functions for text comprehension:

1. decoration, if the pictures only decorate the text;
2. representation, if the pictures give a representation of the text concepts;
3. organisation, if the pictures organise the information conveyed by the text, e.g., via diagrams;
4. interpretation, if the pictures interpret difficult passages or concepts;
5. transformation, if the pictures transform and alter the text meaning, usually to aid recall.

Table 1. recaps such types of functions, with their advantages, disadvantages and the effect on text learning as analysed in [5, 8].

Table 1 The Levie types of functions of images for text comprehension, and their analysed effects on text learning

Function type	Description	Effects on text learning
Decoration	no direct connection with the prose content	null
Representation	telling exactly the same story as the words; overlapping substantially with the text	moderate
Organisation	providing an organisational framework, e.g., diagrams	moderate to substantial
Interpretation	visually interpreting difficult-to-understand passages and concepts	moderate to substantial
Transformation	codifying difficult textual information into a visual metaphoric format that should easy the memorisation of the difficult textual information	substantial

In particular, according to the analysis of Levie et al., pictures that complement the text information being presented increase the likelihood for retention and recall of that information, and hence have substantial effect on learning text. On the other hand, decorative images, not related to the text contents, tend to have no effect on retention and no prose-learning facilitation is to be expected

Starting from the work of Leive et al., Marsh and Domas [7] identify, organise and integrate other functions of images for text comprehension. In their work, images are grouped according to the degree of relation they bear with the text:

1. images that have no relation with the text, e.g., for decoration,
2. images that bear a close relation with the text, e.g., for representation, see Fig. 1,
3. images that interpret or transform the text, e.g., for organisation, interpretation, or transformation, see Figg. 2, 3.

Fig. 1 An example of a representational image from [16]

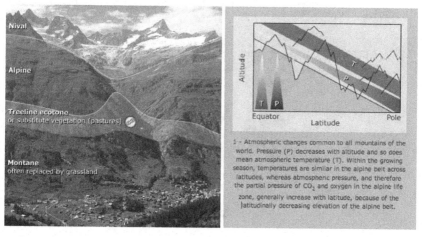

Fig. 2 An organisational pic (on the left) and an interpretational image (on the right) from [16]

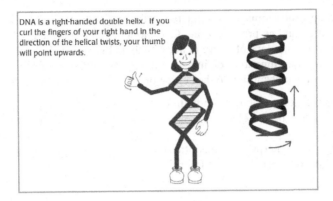

Fig. 3 An example of a transformational image from [16]

The result is a taxonomy of functions of images for text that can be used for analysing the way that images interact with text. It is applicable to all subject areas and all types of documents. We present the taxonomy in Fig. 4, from which we can derive the five functions of images with respect to text analysed by Levie et al. For instance, the decorate concept in the taxonomy in Fig. 4 can be associated to the decorative function of Levie et al; reiterate can be associated to the representation function; organise, interpret and transform can be associated to the Levie et al. functions of organisation, interpretation and transformation.

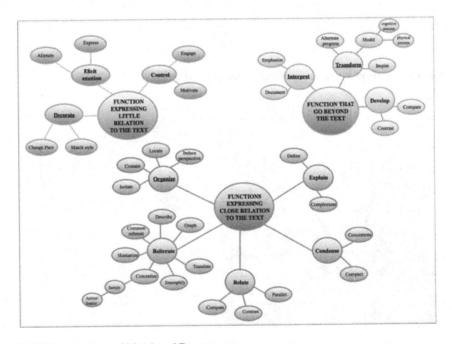

Fig. 4 The taxonomy of Marsh and Domas.

In addition, the taxonomy can be used for assessing the use of images comparatively across documents, for example, assessing the functions played by illustrations in scientific texts and perhaps relating the functions to the effectiveness of the texts for learning or retention of information.

3 Visual Representations for Text Comprehension and Poor Comprehenders

According to the dual coding theory [2, 3], learning is improved when the information is referentially processed through two channels: one for verbal information such as text or audio, the other for nonverbal information such as illustrations and sounds in the environment. Successful readers do this dual coding automatically, for instance through mental imagery. Mental imagery is the process of creating images "in one's mind" while reading. A good deal of research on mental imagery demonstrates that learning of text is enhanced when students are prompted or taught to use mental imagery. In particular, training poor comprehenders to mental imagery by representing specific information in narratives via specific types of visual representations, as good comprehenders do, can be a way to improve the poor comprehenders understanding of narratives [14, 12]. Not surprisingly, perhaps, mental imagery also aided in differentiating good from poor readers.

Table 2 lists several studies concerning training children to mental imagery, aged 7 to 10 years. The table reports the participants in the test, the material used, the training and assessment methods, the reported effects on comprehension. The following acronyms are used in the table and hereafter in the paper:

o   GC: good comprehenders;
o   PC: poor comprehenders;
o   TD: typically developing children;
o   SLI: specific language impairment.

Table 2 Recap of experimental studies concerning mental imagery with children

Users	Age	Author	Material	Training	Assessment	Comprehension
GC vs PC	8	Pressley (1976) [13]	Representational images, text	Physical pictures; sentences with major elements of the text	Questions	Substantial improvement to answering questions

Table 1 (*continued*)

GC vs PC	9-10	Oakhill & Patel (1991) [14]	Tranformational images	Mental pictures	Questions: factual, descriptive, inferential	Imagery group: GC (25) PC (14) Control group: GC (24) PC (13) Substantial improvement on factual and inferential question
GC vs PC	9-10	Gambrell& Bales (1986) [12]	3 high imagery, 2 high-imagery paragraphs, 4 short expository passages, 10-item probing instrument	Two passages with explicit and implicit inconsistency; instruction to induce mental imagery	Questions concerning inconsistencies	Detect the implicit inconsistency: imagery group, 65%; control group, 29%. Control group failed to detect the explicit inconsistencies (71%) and the implicit inconsistencies (73%)
GC vs PC	9-10	Gambrell& Jawitz, (1993) [11]	Representational images. Text both with illustration and not	Mental imagery and illustration Only illustration Only mental imagery	Free recall; clue recall questions	Mental images and illustrations may similarly affect cued recall. The combined strategy use results in deeper processing than is achieved with single strategy use.
TD vs SLI	9	Joffe, Cain & Maric (2007) [15]	Stories presented to the children both verbally and visually; written questions	Form representational images	Questions: factual, inferential	Improvement of 125% in factual; 47% in inferential Pre test factual: SLI (7,5) TD (17) — Post test inferential: SLI (18) TD (23) Pre test Inferent.: SLI (8) TD (22) — Pre test Inferent.: SLI (12) TD (20)

In brief, as exemplified by Cain in [15], the training to mental imagery is an effective way to boost the story comprehension of children with reading impairments. The mental imagery training helped such children to answer questions about short narratives. The improvement was greatest for so-called factual w-questions, that is, concerning facts explicitly narrated in the story. Mental imagery aided the children's memory of explicit details in the text. Moreover, as Cain speculates, the training to the creation of mental images to support text comprehension is immediate and unobtrusive and therefore may be more readily accessible to and accepted by children in the classroom context.

4 Discussion and Conclusions

According to our review of the literature, briefly sketched above, as well as brainstorming meetings conducted with experts of poor comprehenders within the TERENCE EU project and with experts of pedagogy within the DARE project, there are currently no studies concerning *specific* visual representations of narrative events and their causal-temporal relations for poor comprehenders. In spite of this, there are lessons one can learn from the current evidence in the literature, overviewed in this paper, which suggests the direction to move along for designing specific visual representations of events and their relations, aiding their comprehension, for the TERENCE and DARE projects.

First of all, as reported in Section 3, the mental imagery *training* through representational or transformational images significantly helped readers with specific language impairments, like poor comprehenders, to answer factual w-questions concerning events of short narratives. This can be immediately taken over by DARE and TERENCE for designing interventions focusing on features of the narrated events, e.g., who is involved in a specific event narrated in a story.

The lack of a significant impact on inferential processing of this study does not detract from the finding that a relatively short period of training boosted the recall of content dramatically for those struggling readers. Possibly, this lack of impact may also be due to the fact that the training was often done with representational or transformational images. Based on this and the empirical review of [5], we speculate that *organisational* and *interpretative* visual representations, often used in children science textbooks for representing causal or causal-temporal relations, may aid deep text comprehension and, in particular, inference-making of information concerning causal-temporal relations in a narrative. Therefore, the next step of our work foresees the design of visual organisational and interpretational visualisations of causal-temporal relations, based on metaphors used in science textbooks for children, within DARE, and hence their development within TERENCE, The subsequent step of our work is the user-centred-design evaluation of the representations with experts of the domain, and then with 8-11 old poor and good comprehenders in small scale studies.

Moreover and more in general, the training to the creation of mental images for supporting text comprehension is immediate and unobtrusive and therefore may be more readily accepted by poor comprehenders in an educational context, as envisioned in the TERENCE project.

Acknowledgments. The authors' work was supported by TERENCE project, funded by the EC through the FP7 for RTD, Strategic Objective ICT-2009.4.2, ICT, TEL. The contents of the paper reflects only the authors' view and the EC is not liable for it. The second and third authors' work was also funded through the DARE project, financed by the Province of Bozen-Bolzano.

References

1. Najjar, L.J.: Dual coding as a possible explanation for the effects of multimedia on learning (GIT-GVU-95-29). Georgia Institute of Technology, Graphics, Visualization and Usability Center, Atlanta (1995a)
2. Paivio, A.: Dual coding theory: Retrospect and current status. Canadian Journal of Psychology 45, 255–287 (1991)
3. Paivio, A., Csapo, K.: Picture superiority in free recall: Imagery or dual coding? Cognitive Psychology 5, 176–206 (1973)
4. Levie, W.H.: Research on pictures: A guide to the literature. In: Willows, D.M., Houghton, H.A. (eds.) The Psychology of Illustration. Basic Research, vol. 1, pp. 1–49. Springer (1987)
5. Levin, J., Anglin, G., Carrney, R.: On empirically validating functions of pictures in prose. In: Willows, D.M., Houghton, H. (eds.) The Psychology of Illustration. Basic Research, vol. 1, pp. 51–85. Springer (1987)
6. Levie, W.H., Lentz, R.: Effects of text illustrations: A review of research. Educational Communications and Technology Journal 30(4), 195–232 (1982)
7. Marsh, E., Domas White, M.: A taxonomy of relationships between images and text. Journal of Documentation 59(6), 647–672 (2003)
8. Carney, R., Levin, J.R.: Pictorial illustrations still improve students learning from text. Educational Psychology Review (2002)
9. Marschark, M., Sapere, P., Convertino, C., Mayer, C. W.L., Sarchet, T.: Are Deaf Students' Reading Challenges Really About Reading? (in Press)
10. Gambrell, L.B., Jawitz, P.B.: Mental imagery, text illustrations, and children's story comprehension and recall. Reading Research Quarterly 28, 264–276 (1993)
11. Gambrell, L.B., Bales, R.J.: Mental imagery and the comprehension-monitoring performance of fourth- and fifth-grade poor readers. Reading Research Quarterly 21, 454–464 (1986)
12. Pressley, G.M.: Mental imagery helps eight-year-olds remember what they read. Journal of Educational Psychology 68, 355–359 (1976)
13. Oakhill, J., Patel, S.: Can imagery training help children who have comprehension difficulties? Journal of Research in Reading 14, 106–115 (1991)
14. Joffe, V.L., Cain, K., Maric, N.: Comprehension problems in children with specific language impairment: does mental imagery training help? International Journal of Language and Communication Disorders 42, 648–664 (2007)
15. Hibbing, A.N., Rankin-Erikson, J.L.: A picture is worth a thousand words: Using visual images to improve comprehension for middle school struggling readers. Reading Teacher 56(8), 758–771 (2003)
16. Images, http://www.dnaftb.org/

Using Virtual Worlds and Sloodle to Develop Educative Applications

David Griol and José Manuel Molina

Abstract. Education is one of the most interesting applications of virtual worlds, as they can create opportunities to offer educative contents with the advantages of online courses, with the feel of "presence" that this immersive environments can provide. While most of social networking resources are mainly focused on sharing contents using a traditional web interface, virtual worlds facilitate the creation of social networks that enhance the perception and communication among its users through the use of additional modalities. In this paper we analyze the main resources provided by the Second Life virtual world and Sloodle to develop educational environments and describe their application in a educative project at the Universidad Carlos III de Madrid.

Keywords: Virtual Worlds, Second Life, Sloodle, E-Learning.

1 Introduction

Social networking has emerged in the context of the Web 2.0 as a global consumer phenomenon in recent years. According to [1], two-thirds of Internet users visit social networking or blogging, an activity that consumes 25% of the time spent in the network. In fact, the access to social networks is currently the most used Internet activity using both computers or mobile devices to access Internet. The importance acquired by this type of activity is not only profoundly changing the ways of communication, share information and interact with Internet users, but also has a great impact in our daily lives [2, 3].

The development of so-called Web 2.0 has also made possible the introduction of a number of Internet applications into many users' lives, which are profoundly

David Griol · José Manuel Molina
Group of Applied Artificial Intelligence (GIAA), Computer Science Department,
Universidad Carlos III de Madrid
e-mail: {david.griol,josemanuel.molina}@uc3m.es

P. Vittorini et al. (Eds.): International Workshop on Evidence-Based TEL, AISC 152, pp. 99–106.
springerlink.com   © Springer-Verlag Berlin Heidelberg 2012

changing the roots of society by creating new ways of communication and coop-
eration. The popularity of these technologies and applications has produced a con-
siderable progress over the last decade in the development of social networks increasingly complex. Among them, we highlight virtual social worlds, which are
computer-simulated graphic environments in which humans, through their avatars,
"cohabit" with other users. Although traditional virtual worlds were structured a
priori predefining tasks performed by their users, social interaction has currently
a key role in these environments and users can determine their experiences in the
virtual world according to their own decisions. Thanks to the social potential of virtual worlds, they are becoming a useful tool in the teaching-learning process [4, 5].
This way, virtual environments currently enable the creation of learning activities
that provide an interactivity degree that is often difficult to achieve in a traditional
classroom, encouraging students to become protagonists of the learning process and
also enjoy while they are learning.

However, most of the virtual campus and educational applications in these immersive environments have only been created to replicate real world places without
providing benefits from, for instance, consulting these applications in a classical
webpage. This way, several initiatives and research projects currently focus on the
integration of virtual worlds and virtual learning environments. One of the most important initiatives is Sloodle (Simulation Linked Object Oriented Dynamic Learning
Environment) (*http://www.sloodle.org*)), a free and open source project which integrates the multi-user virtual environments of Second Life (*http://secondlife.com/*)
with the Moodle learning-management system (*http://moodle.org/*). This way, Sloodle provides a range of tools for supporting learning and teaching to the immersive
virtual world which are fully integrated with the Moodle web-based learning management system, currently used and tested by hundreds of thousands of educators
and students worldwide.

In this paper we summarize our experience in a Teaching Innovation Program at
the Universidad Carlos III de Madrid, in which we have defined the main objective
of studying a set of basic tools and utilities that are provided by the Second Life
virtual world in combination with Sloodle to develop educational applications. In
addition, we describe the application of these utilities in a experience with students
of the Computer Engineering Degree at our university.

2 Educative Applications of Virtual Worlds

As stated in the introduction, virtual words provide a combination of simulation
tools, sense of immersion and opportunities for communication and collaboration
that have a great potential for their application in education. However, as criticized
in [6], many of the existing educative experiences in virtual worlds only replicated
traditional approaches into the new environment, such as for example recreating
classrooms co-located in a virtual world. Although such direct translation does not
leverage all the potential of the technology, it provides an added value in the case

of online courses, for example, as a mean of e-assessment for large classes during a course.

In these cases, virtual worlds have been usually employed to replicate real world activities, and it has been reported that experiences in the virtual world have a similar effectiveness in terms of learning than in real world [7]. Nevertheless, field experience is usually needed to obtain more meaningful learning. Anyhow, virtual settings that reproduce real situations help to reduce costs or to offer improved learning activities to the students with the same budget. For example, [5] used avatar-mediated training in medicine to teach the students how to deliver bad news.

Virtual environments also allow the creation of learning activities that cannot take place in the real world, thus fostering experiential learning or "learning by doing" [8]. In addition, virtual worlds allow creating manageable representations of abstract entities and thus help students to construct mental models by direct observation and experimentation. For instance, Mikropoulos et al. identified that science and technology courses had much more presence in virtual worlds than social studies and argued that this might be because the concepts explained are usually more abstract, unobservable or far from everyday experience [4].

Other interesting unique characteristics of virtual worlds which were not exploited in these cases are immersion, manipulability and first order experience. In some virtual environments students are immersed in the virtual world, this provides and enhanced interactivity which would be very difficult in a traditional classroom and that makes the students become protagonists of the learning experience [9]. Such a sense of immersion generally fosters engagement of the students with the tasks and the course, or, in a more general sense, with the formation of their own understanding. According to [4], most of the educational studies of virtual worlds claim that virtual interaction favors engagement.

Virtual worlds also provide new opportunities for collaboration, as they facilitate synchronous and asynchronous communication, supporting and enhancing student-student and teacher-student interaction. Petrakou reported that the students involved in their research socialized just like they do in real world courses, trying to get to know each other, and also learning to cope with the new environment [10]. They also reported that the student-student interaction continued after the course when some students discussed the assignments and tried to reach a common understanding.

To fully benefit from all the described characteristics of the virtual worlds, the course contents must be active and designed following a pedagogical strategy that leverages the potential that virtual worlds offer [7]. This way, knowledge is acquired on the basis of authentic real-world problems, for which the solution is a communal task that must be achieved as the union of the individual efforts in the virtual world.

3 Educational Applications Using Second Life and Sloodle

Second Life (SL) is a three dimensional virtual world developed by Linden Lab in 2003 and accessible via the Internet. SL is one of the most popular virtual social worlds: a population today of millions of residents worldwide. It uses very advanced technologies for the development of realistic simulations, so that avatars and the environment are more credible and similar to real world users. The ability to customize SL is extensive and facilitates innovation and student participation, which also enhances the naturalness of the interactions that take place in the virtual world. SL is currently being used with success as a platform for education by many institutions, such as colleges, universities, libraries and government entities.

Besides the previously described features, Second Life presents several utilities specifically tailored for their use in education; probably the most relevant is Sloodle, an open source project which integrates Second Life with the Moodle learning-management system. Sloodle provides a range of tools for supporting learning and teaching to the immersive virtual world. Firstly, it allows controlling the user registration and participation in a course thanks to an *Access Checker* as depicted in Figure 1. Also new users can register in a course using the so-called *Enrol_Booth*.

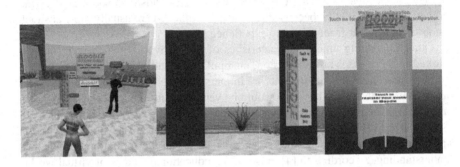

Fig. 1 *Sloodle Access Checker* activated by entering a login zone (left), *Access Checker* with a door (middle) and *Enrol_Booth* (right)

Secondly, there are several tools to create surveys in Sloodle, such as *Choice Horizontal*, *Quiz Chair* or *Quiz Pile* on (Figure 2). *Choice horizontal* allows instructors to create and show surveys in the virtual world, compile the information and show the results in a course. With *Quiz Chair* an avatar can answer questionnaires of a course in SL, while with *Quiz Pile On* provides a similar functionality with a more amusing format. In *Quiz Pile On*, questions in the form a text that floats over a pile, the students seat over the correct answer and if the answer is wrong, he falls over.

Thirdly, the *Sloodle Presenter tool* (shown in Figure 3) allows creating presentations in Second Life, which can combine images, web pages and videos and may be configured so that any avatar or only the owner of the corresponding sim controls the display of the presentation. Finally, there are other interesting tools for object sharing such as the *PrimDrop*, which allows students to deliver their works by sending objects in Second Life, or the *Vending Machine*, which can be used to deliver object to the students (Figure 3).

Fig. 2 *Choice Horizontal*, *Quiz Chair*, and *Quiz Pile* On utilities

Fig. 3 *PrimDrop* (left) and *Vending Machine* (right)

4 Practical Educative Experience

One of the main objectives defined in a Teaching Innovation Program in which we participate at the Universidad Carlos III de Madrid is the development of enhanced e-learning systems to accelerate the learning process, facilitate access, personalize the learning process, and supply a richer learning environment. This way, the study and application of the different tools and utilities described in the previous section was defined as a key point to fulfill this objective. The subject Compilers of the Computer Science Degree was selected for the elaboration of different questionnaires for the experience. The methodology defined for the evaluation of this subject, promoted by the introduction of the new European degrees at our university, emphasizes students' continuous assessment based on students' effort and active participation in their learning. The application of this experience is defined as training activities and also as a part of the continuous assessment of the subject. In terms of training activity, the program aims to adequately follow not only the student learning process throughout the delivery of the proposed activities but also to reinforce students about the assimilation degree of the learning objectives and competencies acquired during the course.

A set of questionnaires with a total of 110 questions, practical cases and problems was elaborated for the subject to carry out the study and evaluation of the educational utilities provided by SL Sloodle. A total of 119 students, in groups of

40 students, participated in the experience. For the generation of the different questions and practical cases for the subject, we considered the following types of exercises:

- Questions concerning theoretical contents as a review of methodologies and concepts (e.g., what is a token?, which are the main modules that make up a compiler? which are the main methodologies studied to develop a syntax analyzer?);
- connection with programs (like the Flex (*http://flex.sourceforge.net/*) lexical analyzer generator and the Yacc (*http://sourceforge.net/projects/byacc/*) parser generator. These programs were used to propose students to deal with practical implementations and provide them with code execution (e.g., define the regular expressions to detect identifiers and numbers in a specific programming language, write a parsing grammar to detect valid if-then-else statements);
- practical cases proposed to the student to obtain conclusions about the appropriate processes for resolve specific problems (develop a practical compiler given the requirements of a simple programming language, test the application of the SLR(1) technique to parse a given input statement, apply the described optimization techniques to reduce the cost execution of a specific programming code).

These questionnaires were implemented by means of a repository for the questions in the Moodle space at the University and using the tools provided by Sloodle for viewing and implementing them in the Second Life virtual world. It should be emphasized that the total of questions were answered by 89% of students. A 86% percentage of students expressed the usefulness of the provided cases and problems to facilitate the achievement of the objectives of the course, enhance their learning and facilitate knowing the degree of understanding of the different contents.

The experience was planned so that in an initial session students were introduced to the concepts needed to interact in the virtual world (download and installation of the browser, creation and personalization of the avatar, resources to search and locate resources, teleportation, use of the described educational utilities, etc). Then, a laboratory session with a duration of 90 minutes was carried out at the end of each one of the units of the course.

In these sessions students accessed the virtual world to answer the questionnaires elaborated with the contents corresponding to each unit in the subject, so that each session served to reinforce the learning process acquired with the master classes. In addition, the participation of a lecturer in the coordination of each one of these sessions made possible the detection of the concepts for which there was a greater number of errors in completing the questionnaires. Thus, each of the laboratory sessions with Second Life was followed with a master lecture to address identified problems and reinforce these concepts.

We have already completed a preliminary evaluation of the experience based on a questionnaire to assess the students' opinion about their previous knowledge on the used technologies, value the possibilities of interaction and communication in the virtual world, assess the the educative utilities proposed in the different activities, and evaluate the learning process. The questionnaire included the following 10 questions:

- *Q1*: State on a scale from 1 to 5 your previous knowledge about new technologies for information access;
- *Q2*: How many times have you accessed virtual worlds like Second Life?;
- *Q3*: Was it easy for you to interact with the different utilities to complete the questionnaires?;
- *Q4*: Do you think that the design of the learning environment is correct?;
- *Q5*: Do you think that the questionnaires cover the main contents of the course?;
- *Q6*: Was it easy for you to communicate with the other participants?;
- *Q7*: Were you sure about what to do at every moment?;
- *Q8*: Do you think that the experience has helped you to better prepare the course?;
- *Q9*: In general terms, are you satisfied with the experience?;
- *Q10*: Indicate which are the most valuable points of the experience and what you should improve.

The first nine questions were assigned a numeric value between one and five (in the same order as they appear in the questionnaire). Table 1 shows the average, minimal and maximum values for the subjective evaluation.

Table 1 Results of the preliminary evaluation of the educative experience (1=minimal value, 5=maximum value)

	Q1	Q2	Q3	Q4	Q5	Q6	Q7	Q8	Q9
Average value	4.6	2.8	4.3	3.8	4.6	3.1	3.5	4.5	4.4
Maximum value	5	4	5	5	5	4	4	5	5
Minimal value	4	1	3	3	3	3	2	4	4

From the results of the evaluation, it can be observed that students positively evaluated the most relevant aspects of the experience, such as the quality of the proposed utilities and contents designed to develop of the learning environment, the potential of virtual worlds to the development of educative activities and the possibilities of communication, socialization and interaction that these immersive environments provide. In addition, they highly appreciated the experience in terms of the main objectives of facilitating the learning process and reinforcing the acquisition of the contents. The set of points to be improved included the possibility of extending the number of activities, the facilitation of a more detailed feedback for each incorrectly answered question, as well as the doubts originated with the use of some of the described utilities. Students also emphasized the facility of answering the required questions, learning from the provided feedback and also enjoying and socializing with the rest of students thanks to the functionalities provided by the Sloodle tools and the interaction in the Second Life virtual world.

5 Conclusions

Social networks and virtual worlds offer a wide range of educational opportunities that make them immersive learning scenarios in which students can explore, meet other residents, socialize, participate in individual and group activities, and cooperate to create the environment. This contribution is focused on presenting the results of a study of the educational utilities provided by the Sloodle tool, which makes possible the use of Moodle in the Second Life virtual world. To do this, we have applied these tools in the learning process of one our the subjects. The results of the project show both the good evaluation by students as well as the educational potential of these tools. As future work, we are extending the experience during this academic year, including in our study several functionalities to adapt the environment and Sloodle tools taking into account student's specific needs, considering their evolution during the course as one of the main aspects to perform this adaptation. We also want to evaluate the benefits of using the Sloodle tools in combination with other virtual worlds like the ones that can be created using OpenSimulator (*http://opensimulator.org*).

Acknowledgements. Research funded by projects CICYT TIN2011-28620-C02-01, CICYT TEC 2011-28626-C02-02, CAM CONTEXTS (S2009/TIC-1485), and DPS2008-07029-C02-02.

References

1. Nielsen: Global Faces and Networked Places: A Nielsen Report on Social Networking's New Global Footprint. Nielsen Online (2009)
2. Boyd, D., Ellison, N.: Social Network Sites, Definition, History and Scholarship. Journal of Computer Mediated Communication 13(1), 210–230 (2007)
3. Dwyer, C.: Digital Relationships in the 'MySpace' Generation: Results from a Qualitative Study. In: Proc. of HICSS 2007, pp. 19–28 (2007)
4. Mikropoulos, T., Natsis, A.: Educational virtual environments: A ten-year review of empirical research (1999-2009). Computers & Education 56(3), 769–780 (2011)
5. Andrade, A., Bagri, A., Zaw, K., Roos, B., Ruiz, J.: Avatar-mediated training in the delivery of bad news in a virtual world. Journal of Palliative Medicine 13(12), 1–14 (2010)
6. Girvan, C., Savage, T.: Identifying an appropriate pedagogy for virtual worlds: A Communal Constructivism case study. Computers & Education 55(1), 342–349 (2010)
7. Jarmon, L., Traphagan, T., Mayrath, M., Trivedi, A.: Virtual world teaching, experiential learning, and assessment: An interdisciplinary communication course in Second Life. Computers & Education 53(1), 169–182 (2009)
8. Ellison, K., Matthews, C.: Virtual history: A socially networked pedagogy of Enlightenment. Educational Research 52(3), 297–307 (2010)
9. Bailenson, J., Yee, N., Blascovich, J., Beall, A., Lundblad, N., Jin, M.: The Use of Immersive Virtual Reality in the Learning Sciences: Digital Transformations of Teachers, Students, and Social Context. Journal of the Learning Sciences 17(1), 102–141 (2008)
10. Petrakou, A.: Interacting through avatars: Virtual worlds as a context for online education. Computers & Education 54(4), 1020–1027 (2010)

The User Classes Building Process in a TEL Project

Tania Di Mascio, Rosella Gennari, Alessandra Melonio, and Pierpaolo Vittorini

Abstract. Nowadays, circa 10% of 7-11 olds turn out to be poor comprehenders: they demonstrate text comprehension difficulties, related to inference making, despite proficiency in low-level cognitive skills like word reading. To improve the reading comprehension of these children, TERENCE, a technology enhanced learning project, aims at stimulating inference-making about stories. In order to design and develop the TERENCE system, we use a user centred design approach that requires an in depth study of the system's main end-users, namely, its learners and educators. This paper reports the user classes building process for learners by means of user-centred design field studies.

1 Introduction

Text comprehension skills and strategies develop enormously from the age of 7–8 until the age of 11, when children develop as independent readers. Nowadays, more and more novice comprehenders turn out to be poor (text) comprehenders: they demonstrate difficulties in deep text comprehension, despite well developed low-level cognitive skills. These poor comprehenders represent the end-users type we refer to in this paper.

Finding stories and educational material that are appropriate for is a challenge, and hence teachers are left alone in their daily interaction with them. Most reading material for 7-11 old children is paper based, and is not easily customisable to the specific requirements of poor comprehenders, e.g., in the types, number or position of temporal connectives. Few systems promote general reading interventions

Tania Di Mascio
University of L'Aquila, DIEI, Via G. Gronchi 18, 67100 L'Aquila, Italy

Rosella Gennari · Alessandra Melonio
Free University of Bozen, Fac. of Computer Science, Pz. Domenicani 3, 39100 Bolzano, Italy

Pierpaolo Vittorini
University of L'Aquila, MISP, Via S. Salvatore, 67100 L'Aquila, fraz. Coppito, Italy

P. Vittorini et al. (Eds.): International Workshop on Evidence-Based TEL, AISC 152, pp. 107–114.
springerlink.com   © Springer-Verlag Berlin Heidelberg 2012

(see e.g., [3]), but they have high school or university textbooks as reading material, instead of stories, and are developed for old children or adults, and not specifically for 7–11 old poor comprehenders.

TERENCE is an FP7 ICT TEL project for 7–11 year old poor comprehenders, hearing and deaf, and their educators, and it aims at filling such a gap: its reading material (in English and in Italian) will be stories adapted to the specific requirements of such poor comprehenders, and its reading interventions are mainly smart games that stimulate inference-making on text, fostering deep text comprehension. In particular, TERENCE builds an *adaptive learning system* (ALS) for improving the reading skills of 7–11 old poor comprehenders, hearing and deaf. The conceptual model of the TERENCE ALS is described in [1], [2]. In particular, the domain model structures the material of the ALS, namely, stories and games. The user model structures the information concerning the end users of the system, that are: *learners*, that is, 7–11 old children; *educators*, namely, teachers and parents; *experts*, that is, end users providing material for the system.

To guide the design and development of the TERENCE ALS, we adopted the *user centered design* (UCD) methodology [5], which involves the end-users into the project from the very beginning, and aims at the overall usability of the system. In an UCD approach, it is custom to classify potential users of the system under design according to the expected usage of the system. It is important to underline the difference between *user types* and *user classes* here: the first are based on the designer's hypothesis about the users based on the literature and/or documentation studies. Such a hypothesis is generally a reasonable assumption, however, it may not fully reflect the specific context of use and its users, since it based on literature rather than on actual research of the specific target users for the context of use analysis. User classes, on the other hand, are based on the analysis of the context of use. The classification is not a hypothesis of the designers, but the result of the interaction between the designers and the users involved in the context of use analysis.

The goal of this paper is to provide a description of the user classes building process for the learner type (see Sect. 2) coming from field studies conducted in Italy (as described in [8] and [10]) on the one hand, and, on the other hand, on the input from other sources (e.g., brainstorming meetings, documents, see [9]). The paper ends with Sect. 3 focused on future work.

2 The User Classes Building Process

This section contains a description of the building process of user classes for the learners' type in the TERENCE project. It starts with row data (e.g., personal data) and ends with a classification of such users. This building process is based oh the field study conducted in Italy in the middle of 2011, in turn based on field studies run at the beginning of 2011. While the first field study, described in [8], focused on knowing our users and their main needs in relation to the overall goal of the project, the second field study focused on defining the classes of users representing the starting point of the SW engineering process of the TERENCE ALS. Following

UCD practices, see e.g., [6] and [7], we conducted experiments using user-based criteria [9]. The experiment design is reported in Subss. 2.1, the user description is reported in Subss. 2.2, the user teaching is reported in Subss. 2.3, the experiment execution is reported in Subss. 2.4, and the result analysis is reported in Subss. 2.5.

2.1   Experiment Design

The general framework for the design of the user field study is described in [8]. Based on the conceptual maps of this framework, specific user tasks were designed to assess the users' characteristics during the study. See, e.g., Tab. 1.

Table 1 Extracurricular activities.

Goals: To assess the learners' extracurricular activities. **Materials**: Stickers; sheets with pictures. **Description**: Each learner received a sheet with stickers representing activities (e.g., reading, going on the internet, drawing, using an iPhone, running, or going to the park), and a blank sheet with three empty circles representing morning, afternoon and evening. Learners had to paste the relevant stickers into the appropriate circles.	

2.2   User Description

The experiments involved 18 teachers and 282 learners (7-11 years old). Consider that 8 olds belong to Class 3. Tab. 2 presents the user involvement.

2.3   User Teaching

Two of the participant schools, namely, Campalto and Torre di Mosto, had already participated to the first field study. They were already informed about the goals of the TERENCE project. For the other schools, first of all, the experimenters decided

Table 2 An overview of the learners participating in the field study.

School	Class	Deaf/hearing unit	Number of learners
Primary school, Torre di Mosto	3	Hearing	42
Institute Gramsci Campalto	3	Hearing	16
	3	Deaf	1
Pescasseroli	2	Hearing	20
	3	Hearing	21
	4	Hearing	17
	5	Hearing	19
Masseri Avezzano	2	Hearing	35
	3	Hearing	38
	4	Hearing	18
	5	Hearing	16
San demetrio ne' Vestini (AQ)	6	Hearing	37

with headmasters the date of a preliminary meeting to present TERENCE. Once the project was clear to all the involved teachers, practical arrangements were made and general information about the field study tasks were explained, e.g., the duration of class sessions and interviews, the nature of the tasks. During these meeting with teachers, all of them asked the experimenters to administer the same test to all children in the classroom without considering if a child is deaf or hearing, poor or good comprehender. For this reason, where deaf and hearing children are mixed in the same classroom, the experimenters administered the same test asking support teachers to aid deaf children.

2.4 Experiment Execution

The field study sessions were conducted in a period of May, year 2011. In two classes of the San Demetrio ne' Vestini school we conduced the study in two days in the week of May 16th. In 4 classes of the Pescasseroli school, we conduced the study in two different days in the week of May 20th. In three classes in the Torre di Mosto and the Campalto schools, we conduced the study in one day on 26 May. In six classes of the Avezzano school, we conduced the study on May 28th. This resulted in a total of 15 primary school classes with children of all ages. In each class, we spent about 90 minutes with learners, and each class teacher participated in a 20-minutes interview. All tasks described in Sect. 2.1 were done with all learners and we collected all assignments and stored the interviews with teachers.

2.5 Result Analysis

In order to describe user classes, we used the persona framework, see [4]. To do this, we applied the following procedure during our data analysis.

We here report the analysis of data for the learners, gathered conducting a quantitative analysis, described by the following procedure.

1. *Data Management:* all data gathered using the assignments described in the experiment design section (Sect. 2.1) is stored in a database for quantitative analysis.
2. *Statistics:* statistics of the quantitative data were calculated using $Chi2$ and $Fisher's$ analysis. Natural variables like gender and age were defined. Other dichotomy variables were derived from statistics observations (e.g., rural versus urban).
3. *Data Analysis:* graphics describing the variables associations were depicted. Using these data, we derived a first classification that stems from orthogonal dimensions (e.g., North/Centre) and sorts learners according to opposite dimensions, or dichotomies.
4. *User Classes:* using tables, graphics and variables associations, we derived four classes, obtained by excluding some classes based on the behavior of variables associations.

Hereinafter, we detail such steps.

2.5.1 Data Management

All data was stored in an open source DBMS: Open Office. We designed the ER diagram. Attributes of the DBMS were derived from the collected data, reading the learners responses.

The most important tables are the following ones.

- School: this table represents the involved schools. The attributes are:
  - Name: the name of the school;
  - City: the city where the school is located (so we can find out if it belongs to the North or the Centre of Italy);
  - Type: the area where the school is located: rural or urban.

- Learner: this table represents the involved Learners. Attributes are (for example) school, classroom and sex.
- Activities: this table represents all the activities shown in the Extracurricular activities task.
- Learner-has-Activities: this table collects the Learners' responses for the Extracurricular activities task.
- Map: this table collects the Learners' responses for Technology use task and the Games task.
- Homework: this table collects the Learners' responses for Homework task.

2.5.2 Statistics

Afterwards, statistics of the quantitative data were calculated. More specifically, we used the STATA software. Data are presented in relation to the following pure and dichotomy variables:

1. Gender (or Sex) – Male (M) and Female (F);
2. Disability – Hearing (H) and Deaf (D);
3. Skills – Poor Comprehension (PC) and Good Comprehension (GC);
4. Region – North and Centre of Italy (N and C);
5. Area – Rural (R) and Urban (U);
6. Socio-Cultural Level – low level (lS) and medium/high level(hS);
7. Age Class – divided into

   - Low Age: children that are 7 to 8 years old;
   - High Age: children that are 9 to 11 years old.

2.5.3 Data Analysis

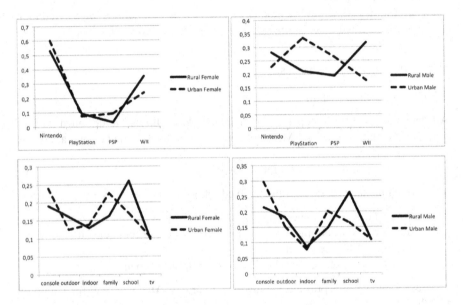

Fig. 1 Cross Analysis – Area vs Gender.

During the statistical analysis, we observed that the significant dichotomy variables with respect to our analysis (e.g., activities, console use and avatar choice) are: (a): Age Class (lowAge and highAge); (b): Gender (M and F); (c): Region (North Italy and Centre of Italy); (d): Area (Rural and Urban).

In order to discover which dichotomy variable is most representative for activities and console use, we analysed the data described in the previous step via cross associations of area, location and age class versus gender, see the Fig. 1.

Observing the graphs in Fig. 1 we observed that both urban female and rural female curves have the same behavior regarding activities and console use. This means that we can decide to not consider the area dichotomy variable when considering female learners. This is not the case when we observe rural and urban male curves. They have the same behavior only in the console use case, but not in the activity case. This means that we will consider area as a dichotomy variable when we derive classes for male learners. We proceeded with depicting the graphs representing the cross analysis between gender and location here not reported for space constraints. Starting from these data we decided to consider location as a dichotomy variable when deriving classes for male learners. We also analysed graphs representing the cross analysis between age class and area. We decided to disregard the area dichotomy variable when considering male as well as female learners. From the cross analysis between gender and age class variables, we decided to disregard the gender dichotomy variable when considering the age class. We proceeded with graphs representing the cross analysis between area and location deciding to disregard the area dichotomy variable when considering the location variable. Finally, in order to deduce the most important dichotomy variables, we analysed graphs for male learners considering the area plus location versus age class variables, Fig. 2. The two graphs depicted in Fig. 2 show that, in the case of males, we must consider the location as well as the area dichotomy variables. In fact, even if in the case of console use (the right part of the figure) the curves have the same behavior, the behavior is not the same, for activities.

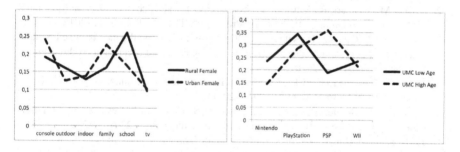

Fig. 2 Cross Analysis – Area and – Location vs Age Class.

2.5.4 User Classes

Studying the behavior of trends in the graphs, we discover that some variables are more significant than others. Based on this, we were able to deduce the following learners classes: (1) – DF (Deaf Female); (2) – HF (Hearing Female); (3) – H R M N lowAge (Hearing Rural Male North lowAge); (4) – H U M N lowAge (Hearing Urban Male Center lowAge); (5) – H U M N highAge (Hearing Urban Male Center highAge).

3   Conclusion and Future Works

In this paper, we described the user classes building process. Currently, we are working on a finer-grained analysis of the user requirements, to be reviewed through small-scale evaluations. The updated requirements will provide the input for the revision of the conceptual model of the TERENCE ALS, in particular, its user model, and hence the implementation of the TERENCE software. Such evaluations, more in general, will serve to assess the usability of our ALS, and in particular: (1) the appeal and adequacy of its learning material, (2) the pedagogical effectiveness of our ALS in improving the text comprehension of 7–11 old poor comprehenders.

Acknowledgments. The authors' work was supported by TERENCE project, funded by the EC through the FP7 for RTD, Strategic Objective ICT-2009.4.2, ICT, TEL. The contents of the paper reflects only the authors' view and the EC is not liable for it. The second author work was also funded through the CRESCO project, financed by the Province of Bozen-Bolzano. The third author work was funded through the DARE project, financed by the Province of Bozen-Bolzano.

References

1. Alrifai, M., Gennari, R., Tifrea, O., Vittorini, P.: The Domain and User Models of the TERENCE Adaptive Learning System. In: Proc. of eb. TEL 2012. Springer (2012)
2. Alrifai, M., Gennari, R., Vittorini, P.: Adapting with Evidence: the Adaptive Model and the Stimulation Plan of TERENCE. In: Proc. of eb. TEL 2012. Springer (2012)
3. Di Mascio, T., Gennari, R., Vittorini, P.: The Design of An Intelligent Adaptive Learning System for Poor Comprehenders. In: Proc. of Cognitive and Metacognitive Educational Systems 2010. IADIS (2010)
4. Gasperis, G.D., Mascio, T.D., Florio, N.: Tatot: a viewer for annotating stories in the terence project. In: Proc. of ITAIS 2011 (2011)
5. Goransson, J., Boivie, B., Blomkvist, I., Persson, S., Cajanger, J.: Key principles for user-centred systems design. Behavior and Information Technology 22(6), 397–409 (2003)
6. Hartson, H., Andre, T., Williges, R.: Criteria for Evaluating Usability Evaluation Methods. International Journal of Human Computer Interaction 13(4), 373–410 (2001)
7. Mascio, T.D., Gennari, R.: A Usability Guide to Intelligent Web Tools for the Literacy of Deaf People. In: Integrating Usability Engineering for Designing the Web Experience: Methodologies and Principles, pp. 201–224. ICI Global (2010)
8. Pasini, M.: Working document 1.1: User classification and identification. Technical report, TERENCE project (2011)
9. Slegers, K., Gennari, R.: State of the Art of Methods for the User Analysis and Description of Context of Use. Technical Report D1.1, TERENCE project (2011)
10. Slegers, K., Mascio, T.D.: Working document 1.2: Usability goals and user needs. Technical report, TERENCE project (2011)

CBR Proposal for Personalizing Educational Content

Ana Gil, Sara Rodríguez, Fernando De la Prieta, Juan F. De Paz,
and Beatriz Martín

Abstract. A major challenge in searching and retrieval digital content is to efficiently find the most suitable for the users. This paper proposes a new approach to filter the educational content retrieved based on Case-Based Reasoning (CBR). AIREH (Architecture for Intelligent Recovery of Educational content in Heterogeneous Environments) is a multi-agent architecture that can search and integrate heterogeneous educational content within the CBR model proposes. The recommendation model and the technologies reported in this research applied to educational content are an example of the potential for personalizing labeled educational content recovered from heterogeneous environments.

Keywords: E-learning, learning objects, Case Base Reasoning, recommender systems, Multi-agent systems.

1 Introduction

The significant evolution in the methods for managing and organizing large volumes of digital content in Internet has led to a phenomenon of globalization and decentralization. The education sector is a significant generator and consumer for digital content placed in large databases in Internet. The main problems to manage all the data is due to the various ways the information is characterized, contained in specialized data repositories each of which must be accessed by their own methods, resulting in dozens of communication protocols. Users have access to technologies that allow them to obtain educational content ubiquitously and in real-time. The information and technological overload become a problem in the search and selection for the educational contents. These aspects placed the current educational environment as a relevant sector in the development and integration of emerging

Ana Gil · Sara Rodríguez · Fernando De la Prieta · Juan F. De Paz · Beatriz Martín
Departamento de Informática y Automática – Facultad de Ciencias
University of Salamanca – Spain
Plaza de la Merced s/n
37008 Salamanca
e-mail: {abg,srg,fer,fcofds,eureka}@usal.es

P. Vittorini et al. (Eds.): International Workshop on Evidence-Based TEL, AISC 152, pp. 115–123.
springerlink.com   © Springer-Verlag Berlin Heidelberg 2012

solutions to personalizing the management processes for the search, retrieval and integration of heterogeneous educational content.

In previous works has been presented an architecture recovery based educational content partner organizations called AIREH [1]. The main novelty was that proposed architecture has the capability of dynamic and adaptive planning to carry out an optimal distribution of the tasks of the organization's member agents enabling intelligent content retrieval and flexibility in the highly dynamic environment for which it have raised. The proposed architecture allows searching multiple repositories simultaneously based on federal search services in the environment of the called Learning Object Repositories (LOR). The LOR contains educational content, called Learning Object (LO) [2], and tagged in any of the educational metadata standards. The architecture AIREH supports mechanisms to implementing the recommendation or ranking for educational content recovered. This paper proposes a new approach to filter the educational content retrieved based on Case-Based Reasoning (CBR) as a basic feature in the active search of educational content within the architecture AIREH.

The remainder of this paper is organized as follows: Section 2 describes relevant works related to educational content recommendation; Section 3 introduces a new approach to applying CBR to the tagged educational content recommendation domain. Selected results of a comprehensive evaluation of the approach are presented in Section 4. The paper closes with relevant conclusions and an outlook to future work in section 5.

2 Related Work

Recommender systems select user information from tastes and preferences on a particular topic. The area of the store was the first field of application of these systems. Nowadays recommender systems are present in all areas of services offered via Internet such as music, news pages or virtual libraries. Its use is also taking a major impact on all types of e-learning systems [3]. This paper seeks to obtain a kind of repository customized for each user based on the effective management of search sequences in different repositories.

There are different methodologies to enable filtering information mechanisms in response to relevant aspects about user and context. To perform this task the system should learn from what the user finds as useful, interesting, or better, for which need to be in contact with the user. In addition to personalized information filtering the mechanism must be flexible to changes in user interests over time by establishing criteria for the treatment of the dynamic content.

The processes of learning and adaptation of recommender systems are based on the discovery of users' pattern through their online transactions or interactions with content. Learning methods are implemented in information filtering systems through automated reasoning mechanisms and learning mechanisms from the field of artificial intelligence mostly. Recommender systems are based on machine learning on the user's previous experiences in context. Among the various technologies for machine learning, this work makes use of Case-Based Reasoning (CBR) as a paradigm for learning and reasoning through experience. CBR models allow the resolution of new situations by adapting solutions and customization through automatic reasoning about solutions to problems given above.

With so many educational content repositories, a major challenge is to efficiently find the most suitable contents for the users. This objective has attracted much research in the field of the selection and recommendation. Researchers and developers of e-learning have begun to apply information retrieval techniques with technologies for recommendation, especially collaborative filtering [4], or web mining [5], for recommending educational content. A recent review of these applications can be seen in [6]. The features that handle these information filtering techniques in this context are the attribute information of education item (content-based approach) and the user context (collaborative approach).

Based on collaborative filtering recommendation algorithms with close neighbors are the first works developed by Altered Vista [7, 8]. These works explore how to collect user reviews of learning resources and propagate them in the form of word-of-mouth recommendations. Others as RACOFI (Rule-Applying Collaborative Filtering) propose a collaborative filtering by rules, by integrating a collaborative filtering algorithm that works with user ratings of a set of rules of inference, which creates an association between the content and rate of recommendation. McCalla [9] have proposed an improvement to collaborative filtering that takes into account the gradual accumulation of information and focus on end users. Manouselis et al. [10] have conducted a case study with data collected from the CELEBRATE portal users to determine an appropriate collaborative filtering algorithm.

Some solutions take a hybrid approach. [11-14] make use of algorithms based on reviews from other users according to interests which are extracted through nearest neighbor algorithms. These correlation-based algorithms are used to calculate an index score on the usefulness of learning objects through the analysis of comments from students with similar profiles. These algorithms improve preference-based selection algorithms by incorporating aspects of student preferences.

The works by [15-18] suggest the need for selecting learning objects by taking into account the educational content described by their metadata. Based on semantic aspects by considering contextual information from the student's cognitive activities and the LO content structure works [19, 20] propose learning objects recommendation to suit the student's cognitive activities through an approach based on ontologies.

There are a growing number of papers proposing systems to recommend learning resources, evidenced by the lack of operational solutions as confirmed by recent work [5]. But it is interesting to highlight that most of the evaluated proposals point out that the incorporation of mechanisms to assess attributes related to the educational content as well as aspects of user context and their interaction with the content, create effective recommendation mechanisms. A closer look at the revised proposals underscores the lack of applications on real systems and educational content. The architecture used in this paper, AIREH, provides multiple perspectives to assess the recovery of educational content from a real, open and scalable environment. The support to actual educational context allows proposing, implementing and making a full study by actual tests on recommendation for recovered LOs based on a CBR.

3 AIREH Framework

It is necessary to provide the user with a framework that unifies the search and retrieval of educational content. AIREH (Architecture for Intelligent Recovery of Educational content in Heterogeneous Environments) is a multi-agent architecture that can search and integrate heterogeneous educational content. AIREH is based on a model with dynamic and adaptive planning capabilities to carry out the optimal distribution of the tasks for the organization's member agents, thus enabling intelligent content retrieval and flexibility in a highly dynamic environment (users, profiles, features, content, variability of learning object repositories', etc.).

This framework, see Figure 1, thus facilitates the learning process based on digital contents because it retrieves and filters learning objects properly while classifies according to rules. The generation of such rules comes from the organization of the content recovered based on educational metadata.

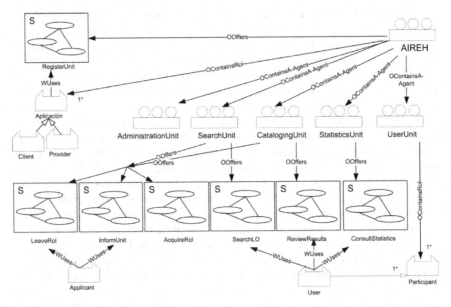

Fig. 1 Organizational model diagram congregational structure (external functionality)

The system offers educational resources as a final product encapsulated in the form of LO. These will be requested by users through a search process. It also provides statistical information such as product performance and use repositories of LO, identifying those that are used according to the search patterns. The mission of the organization will first maximize the system performance of queries by reducing time and increasing performance, and also, maximize the quality of results. The resolution of the issue of content retrieval has been addressed as the federated search mechanism by solving three phases: (1) the selection of repositories, (2) the recovery of content, and finally (3) the merger of results. The sum of several criteria on the ranking of retrieved objects in the system provides a hybrid recommendation that begins with a refined content-based recommendation and collaborative features. This stage ends with the third phase of the federated search, and includes recommendations on merging the content, which improves the quality of the retrieved content for the user that generated the query.

3.1 Overview of the CBR Recommendation Strategy

The Case Base Reasoning (CBR) is a particular search technique widely used in nearest neighbor recommender systems. Recovery techniques and their adaptation to CBR techniques have become effective for the development of recommender systems [21, 22]. This involves the building of user models and the mechanism to anticipate and predict user preferences. In addressing these types of task, recommender systems draw on previous results from machine learning and other AI technology advances.

In order to bring the user with a single ordered list of Learning Objects that incorporates user's relevance criteria this work proposes a CBR reasoning model. In this regard, CBR-BDI agents use a system of case-based reasoning. To facilitate the construction of CBR-BDI agents, different jobs within the works of [22, 23] present a set of equivalence merge the two technologies through a generic library.

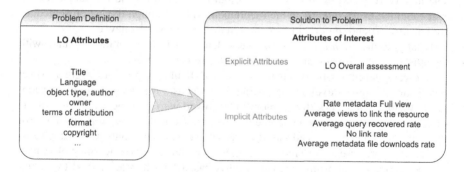

Fig. 2 Example of Case Representation in the educational content domain

A CBR depends largely on the structure and content representation and its collection of cases. The developed system is characterized by working with cases defined the characteristics of the educational context. Each case in the CBR case base is divided into two parts, see Figure 2. The first one describes a set of attributes of the named target (the definition of the problem in CBR terminology). The second element is the set of attributes that describe user interest in the issue because (the solution to the problem through a set of goals in CBR terminology).

With regard to the CBR cycle in the *recovery phase* the system retrieves similar items. Define the degree of similarity between the elements is the most important step in the recovery phase of CBR. The level of similarity between two items is computed using a global function of similarity, calculated by weighting based on weights of various measures of similarity. For numeric attributes, linear similarity functions were designed according to exponential. For attributes labeled, has made use of

WordNet semantic similarity algorithm [24]. With the efficient similarity measure defined, given a new case, we obtain an ordered list of similar cases. Using this feature of CBR, when a user is interested in a subject into the educational context, the recommendation system can generate a list ordered elements based on user profile.

The *reuse phase* consists of adapting the old solutions of retrieved cases for the new problem. Once the system has retrieved a set of above items (most similar), the system knows the user's interest defined by some interest attributes (solution in CBR terminology in Figure 2). Assuming the user interest in a new LO is similar to the interests of another user in similar LO in the reuse phase, the recommendation system calculates a value that encodes the degree of interest for the new element. This value is used to decide the ranking position of the set of LO recovered for the user depending on the grade.

The value of the interest, equation (1), is computed from the values of LO similar interests selected in the recovery phase. We calculate this interest by combining the numerical results assessing the degree of interest for each case in function of explicit and implicit interest.

$$I_c = \mathcal{M}(f^e(Int_j^e), f^i(Int_j^i)) \tag{1}$$

In the *revision phase*, the system evaluates the user's interest in the LO performed. The idea is to track user interaction through the system to find relevant information about the user's interest in the recommended LO through the list presented, as well as explicit and implicit information in order to retain the new case.

Finally, in the *retention phase*, the new element is inserted in the case base with the attributes of interest that were added in the revision phase.

The case definition provided in the CBR-BDI library provides a great power and flexibility to the developer, allowing them to customize the case to his particular problem. In this work the problem to be solved is to rank, according to criteria of customization, educational content retrieved from different repositories. For this initial problem is defined based on the elements of the context for which it is necessary to propose an ordered list of relevant results retrieved learning objects.

The recommendation mechanism used is based on a hybrid method that combines filtering techniques based on collaborative content aspects. This collaborative aspect comes from the feedback of users of the LO and is collected through the interface, and becomes part of the attributes explicit in the case base described in this section.

4 Experimental Results

The recommendation is made by implementing the CBR proposed mechanism and according to the group of recovered cases. To validate the recommendation proposal, we evaluated the results obtained by the AIREH assessment by 6 months with 10 users. They perform a battery of queries from a selection of keywords from the computer science ground extracted from UNESCO codes.

We evaluate the user perception about the quality of the recommendations made by the proposed mechanism along de CBR evolution. The evolution of the number of cases in the case base allows for greater knowledge and appreciation of potential LO. This improvement is due to the system's ability to learn and adapt to lessons learned. Likewise, the experiences allow a better adaptation to the user profile.

The Figure 3 shows the user perception about the quality of the recommendations made by the proposed mechanism through the interface. Each user has evaluated over 12 months the quality of the learning objects in the first five places in the recommender. Success in the system is evaluated through user interaction with the recommended LO, as well as the assessment it makes of each. The user perceives an improvement in the time of the recommended LO.

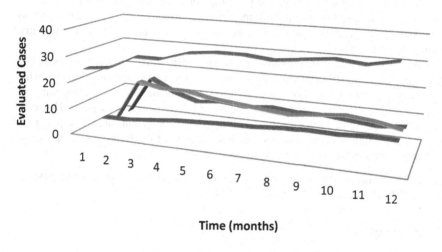

■ Number of Cases ■ Evaluation of the n-LO ■ Base Access ■ Updated Cases

Fig. 3 Evaluation of the recommendations of the CBR

The Figure 3 shows also that the number of updated cases decreases as the system acquires experiences (the X axis represents the evaluations during the time period and the y axis quantifies the number of cases concerning the aspects evaluated (number of cases, access to the base and updates of cases). It is logical, since by increasing the variability of the cases captured and the number, the ability to find cases similar to the query that the user requires increases and may validate the recovered LO criteria defining user tastes and/or needs.

5 Conclusions and Further Work

The new approach for intelligent search of educational content is introduced by the framework prototype called AIREH. This architecture is based on the application of virtual organizations of multi-agent systems and is applied to a federated search in repositories of learning objects and the subsequent recommendation of results. AIREH provides multiple perspectives to assess recovery of educational content from a real, open and scalable environment while the support to implement recommendations or ranking mechanisms for LO recovered.

This work demonstrate the possibility of modeling an efficient system for managing open systems from a model of an adaptive organization that provides personal recovery based on a CBR while is flexible and dynamic. The type of recommendations used in this work lies in a system combining the hybrid type, since the active user is recommended with educational content that other users have taken in the same field of interest, but taking into account the history of the content selected by the user in advance. The model and the technologies presented in this paper are an example of the potential for developing recovery systems for digital content based on the paradigm of virtual organizations of agents. The advantages of the proposed architecture are its flexibility, customization, integrative solution and efficiency.

Currently we have an operative prototype with some of the features implemented. There are still many others featured are increasing the functionality while different perform evaluations are covered for enhancing the performance of searching and retrieval educational content with personalization aspects. However, the research presented in this paper working on a CBR as learning strategy applied to recommendation has provided suitable strategies for the future research towards enhanced content based retrieval systems.

Acknowledgements. This work has been partially supported by the project MICINN TIN 2009-13839-C03-03 and JCyL project SA225A11-2.

References

1. Gil, A.B., De la Prieta, F., Rodríguez, S.: Automatic Learning Object Extraction and Classification in Heterogeneous Environments. In: Pérez, J.B., Corchado, J.M., Moreno, M.N., Julián, V., Mathieu, P., Canada-Bago, J., Ortega, A., Caballero, A.F. (eds.) Highlights in Practical Applications of Agents and Multiagent Systems. AISC, vol. 89, pp. 109–116. Springer, Heidelberg (2011)
2. Chiappe, A., Segovia, Y., Rincon, H.Y.: Toward an instructional design model based on learning objects. Educational Technology Research and Development 55, 671–681 (2007)
3. Manouselis, N., Drachsler, H., Vuorikari, R., Hummel, H., Koper, R.: Recommender Systems in Technology Enhanced Learning. In: Recommender Systems Handbook, pp. 387–415. Springer (2011)
4. Bobadilla, J., Serradilla, F., Hernando, A., Lens, M.: Collaborative filtering adapted to recommender systems of e-learning. Knowledge-Based Systems (2009), doi:10.1016/j.knosys.2009.01.008
5. Khribi, M.K., Jemni, M., Nasraoui, O.: Automatic recommendations for e-learning personalization based on web usage mining techniques and information retrieval. Educational Technology & Society 12(4), 30–42 (2009)
6. Manouselis, N., Vuorikari, R., Van Assche, F.: Collaborative recommendation of e-learning resources: an experimental investigation. Journal of Computer Assisted Learning 26, 227–242 (2010)
7. Recker, M., Walker, A., Lawless, K.: What do you recommend? Implementation and analyses of collaborative information filtering of web resources for education. Instructional Science 31(4-5), 299–316 (2003)

8. Lemire, D., Boley, H., McGrath, S., Ball, M.: Collaborative Filtering and Inference Rules for Context-Aware Learning Object Recommendation. Technolodgy and Smart Education 2(3), 179–188 (2005)
9. McCalla, G.: The Ecological Approach to the Design of E-Learning Environments: Purpose-based Capture and Use of Information about Learners. Journal of Interactive Media in Education (7) (2004); Special Issue on the Educational Semantic Web 1, 18 (2004)
10. Manouselis, N., Vuorikari, R., Van Assche, F.: Simulated Analysis of Collaborative Filtering for Learning Object Recommendation. In: SIRTEL Workshop, EC-TEL (2007)
11. Aijuan, D., Baoying, W.: Domain-based recommendation and retrieval of relevant materials in e-learning. In: IEEE International Workshop on Semantic Computing and Applications 2008 (IWSCA 2008), pp. 103–108 (2008)
12. Ghauth, K., Abdullah, N.: Learning materials recommendation using good learners' ratings and content-based filtering. In: Educational Technology Research and Development, SN 1042–1629. Springer, Boston (2010),
http://dx.doi.org/10.1007/s11423-010-9155-4
13. Tsai, K.H., Chiu, T.K., Lee, M.C., Wang, T.I.: A learning Object Recommendation Model based on the Preference and Ontological Approaches. In: Proceeding of the Sixth International Conference on Advanced Learning Technologies, ICALT 2006 (2006)
14. Wang, T.I., Tsai, K.H., Lee, M.C., Chiu, T.K.: Personalized Learning Objects Recommendation based on the Semantic Aware Discovery and the Learner Preference Pattern. Educational Technology and Society 10(3), 84–105 (2007)
15. Yang, Y.: An evaluation of statistical approaches to text categorization. Journal of Information Retrieval 1, 67–88 (1999)
16. Kerkiri, T., Manitsaris, A., Mavridou, A.: Reputation metadata for recommending personalized e-learning resources. In: Proceedings of the Second International Workshop on Semantic Media Adaptation and Personalization, Uxbridge, pp. 110–115 (2007)
17. Ochoa, X., Duval, E.: Use of Contextualized Attention Metadata for Ranking and Recommending Learning Objects. In: Proceedings of 1st International Workshop on Contextualized Attention Metadata: Collecting, Managing and Exploiting of Rich Usage Information, pp. 9–16 (2006)
18. Wolpers, M., Najjar, J., Duval, E.: Tracking Actual Usage: the Attention Metadata Approach. Educational Technology & Society 10(3), 106–121 (2007)
19. Han, Q., Gao, F., Wang, H.: Ontology-based learning object recommendation for cognitive considerations. In: 8th World Congress on Intelligent Control and Automation (WCICA), July 7-9, pp. 2746–2750 (2010)
20. Ruiz-Iniesta, A., Jiménez-Díaz, G., Gómez-Albarrán, M.: Personalización en Recomendadores Basados en Contenido y su Aplicación a Repositorios de Objetos de Aprendizaje. In: IEEE-RITA, vol. 5(1), pp. 31–38 (2010)
21. Montaner, M., López, B., de la Rosa, J.L.: Opinion-Based Filtering through Trust. In: Klusch, M., Ossowski, S., Shehory, O. (eds.) CIA 2002. LNCS (LNAI), vol. 2446, pp. 164–178. Springer, Heidelberg (2002)
22. Corchado, J.M., Laza, R.: Constructing Deliberative Agents with Case-based Reasoning Technology. International Journal of Intelligent Systems 18(12), 1227–1241 (2003)
23. Glez-Bedia, M., Corchado, J.M., Corchado, E.S., Fyfe, C.: Analytical Model for Constructing Deliberative Agents. Engineering Intelligent Systems 3, 173–185 (2002)
24. Jiang, J.J., Conrath, D.W.: Semantic Similarity Based on Corpus Statistics and Lexical Taxonomy CoRRcmp-lg/9709008 (1997)

Evaluating ZigBee Protocol to Design CAFLA: A Framework to Develop Location-Based Learning Activities

Óscar García, Ricardo Serafín Alonso, Dante Israel Tapia, and Juan Manuel Corchado

Abstract. The inclusion of mobile devices in learning environments has allowed both the emergence of new ways of learning and the adaptation of traditional teaching methods. In this sense, the information about the users and their environment must be taken into account when developing new approaches and solutions. Fortunately, there are several technologies that can help gathering such information. This paper introduces CAFLA, a context-aware framework aimed at helping developers to create location-based learning applications. CAFLA provides tools for integrating several communication technologies that can be used in collaborative learning scenarios.

Keywords: Context-aware Learning, Location-based Learning, Computer Supported Collaborative Learning, Wireless Technologies.

1 Introduction

The impact of technology on our society is difficult to measure, especially when new technologies, devices or services evolve even faster than the users' needs. The advances on personal computers, Internet and mobile devices have changed our world in a political, economic and social way. In this sense, mobile devices are becoming more ubiquitous and usable, allowing new ways of interaction and adding context-aware capabilities for a wide diversity of application scenarios [1].

Education has not stood aside from these advances, allowing the emergence of new ways of learning by improving classic methods through the use of technology or simply by sharing information in a different way. Mobile Learning has become the umbrella under which new ways of learning have emerged, including areas such as *Mobile Computer Supported Collaborative Learning* (MCSCL),

Óscar García · Ricardo Serafín Alonso · Dante Israel Tapia · Juan Manuel Corchado
Department of Computer Science and Automation, University of Salamanca.
Plaza de la Merced, s/n, 37008, Salamanca, Spain
e-mail: {oscgar,ralorin,dantetapia,corchado}@usal.es

P. Vittorini et al. (Eds.): International Workshop on Evidence-Based TEL, AISC 152, pp. 125–132.
springerlink.com   © Springer-Verlag Berlin Heidelberg 2012

based on traditional CSCL, *Context-Aware Pervasive Learning* or, more recently, *Location-Based Learning*. There are several approaches proposed by the scientific community in these research areas which share a common element: the use of mobile devices and wireless communications [2].

This paper aims to identify and characterize the most relevant trends in Mobile Learning through the analysis of relevant work carried out in the research areas mentioned before. This analysis focuses on the study of technologies used to communicate mobile devices, as well as the type of learning activity undertaken. This analysis allows us to introduce the ZigBee communication protocol as a candidate to provide a technology base that enables the development of new ways of learning.

The following section describes the background and problem description related to the approach presented. Then, the main characteristics of the framework proposed are described. Subsequently, the ZigBee communication standard is depicted, focusing on the integration with CAFLA. Finally, the conclusions and future work are presented.

2 Background and Problem Description

The use of mobile devices in educational processes has led to new opportunities for learning [2]. Currently, we are able to interconnect different devices through multiple wireless communication protocols (e.g., Wi-Fi or Bluetootht), to locate them (via GPS) or to collect data from the environment (e.g., using Wireless Sensor Networks). These abilities seem to point to their suitability for different forms of learning.

2.1 Mobile Learning

We intend to focus our study on three specific types of learning that are carried out with mobile devices: Mobile Computer Supported Collaborative Learning, Context-aware Pervasive Learning and Location-based Learning. It is important to clarify that the main reason for choosing these three variants is that they exploit the most important benefits that mobile devices provide: mobility, communication skills including collection and provision of contextual information and precise location anytime.

Mobile Learning is defined as "the processes of coming to know through conversations across multiple contexts amongst people and personal interactive technologies" [3]. This definition implies two important ideas: first of them is that technology can be involved into the learning process; the second idea suggests that mobile learning emphasize the communication between people involved and their interaction with the context [4].

Based on the ability to interconnect devices we can assert that they can be useful to foster collaboration among students, that is, they can act as a tool that supports CSCL (*Computer Supported Collaborative Learning*) [5]. The use of mobile devices into a CSCL system is known as *Mobile CSCL* (MCSCL) [6]. The analysis of the literature allows us to identify a large number of contributions that describe MCSCL systems. Among them we can find modifications of traditional learning management systems that are able to adapt usual utilities from e-learning platforms taking into account the mobile devices' requirements and specifications through its integration into Web Services [7]. Beyond adaptation of traditional e-learning platforms, multiple applications specifically developed to support CSCL using mobile devices have been described. Such applications provide an easier way to improve ubiquitous collaboration or foster face-to-face activities [8].

MCSCL systems are usually designed with a client-server architecture in which all participants join the same network. The introduction of MANETs (*Mobile Ad-hoc NETworks*) into collaborative learning environments with mobile devices is intended to relax this operational model [9]. Moreover, MANETs allow learners to work outside the classroom to enhance collaborative learning both indoor (e.g., museums) or outdoor (e.g., parks) spaces that present any didactic interest [10]. An example of pure ad-hoc network for collaborative learning is the PASIR platform [11], designed to share resources among mobile devices in ad-hoc networks. Its design does not have to consider central elements, as information is fully replicated.

2.2 Context-Aware Learning

Ad-hoc networking is usually related to sensor networks [12]. Together with interconnected and embedded computing devices, auxiliary input/output devices and servers; ad-hoc networking provide a technological base that, when used within the educational environment, allow us to define Context-Aware Learning Environments as "setting in which students can become totally immersed in the learning process" [13]. These environments connect, integrate and share learning collaborators, learning contents and learning services [14]. The most important aspect in this type of system is the data acquisition from the environment in order to customize the learning activities that students must follow [15].

Museum guides especially benefit from this type of system as they can provide information from any area of interest or artwork and customize the content they offer [16]. Those scenarios are environments where users receive a wealth of information from many sources so mobile devices allow receiving real-time contextual information while giving students total freedom of movement to carry out their activities.

2.3 Location-Based Learning

At this time, there are several technologies integrated into mobile devices that can support learning applications, between them these that are related to identification and location. It is easy to realize that the contextual information is determined by two fundamental factors. Firstly, the identification of the element that provides the information. It is important to know anytime who or which offers information and to whom is given so we can customize content basing on the needs of students. RFID identification systems provide a technological base for this proposal [17]. Secondly, the location of the students is a factor to consider when customizing the contextual information: knowing with precision where each person is, we are able to develop activities in which the information received will depend on person's location. Using them, it is possible to seamlessly personalize content to students while they are freely moving throughout the workspace.

In outdoor environments, GPS technology is undoubtedly the most taken by its maturity and good performance. Takacs et al. (2008) describes an example of learning application that uses GPS and augmented reality to provide personalized learning content through mobile phones [18]. However, due to its nature, the GPS system does not work properly indoors. There are alternatives that allow us to have indoor location by means of other technologies, such as RFID [17]. However, the accuracy achieved by this kind of technology does not reach the values obtained with outdoor GPS and the proximity needed between devices becomes an important restriction. In addition, these systems require a tedious, delicate and obtrusive calibration process that affects negatively the learning applications developed over them.

2.4 Synopsis

Throughout this section we have identified different solutions for each of the kinds of learning. Each of the technologies involved in those proposals cover a specific kind of learning. However, it is difficult to use some of them to cover all three kinds of learning at the same time. Joining these learning spaces we would be able to provide a fully adaptable environment to the needs raised by teachers in their design activities. Thus, we can include locating and tracking of objects of interest and students in the learning environment. Then, identifying uniquely them, and recording all the interactions with each other and with the objects, we will be able to generate a ubiquitous learning environment in which fit many activities and subsequent analysis tools that will improve the whole process.

The following section discusses the use of ZigBee as an enabling technology to provide innovative learning scenarios. Its ability to create ad-hoc networks and integrate with wireless sensor networks, as well as the facilities it provides to develop real-time locating systems, allow us to consider this communication protocol as the optimal to cover simultaneously collaborative, context-aware and location-based learning scenarios.

3 CAFLA: Context-Aware Framework for Learning Applications

The use of mobile devices in learning activities implies the use of wireless communication protocols. The best known are Bluetooth, Wi-Fi and GSM/GPRS. If we try to be able to cover the three learning environments presented simultaneously, it would be necessary to use two or more different technologies.

CAFLA (*Context-Aware Framework for Learning Applications*) is a new framework that pretends help developers to easily deploy new learning activities. Its most innovate characteristic is that it covers three kinds of learning that are the most used in mobile learning: Context-Aware Leaning, Collaborative Learning and Location-based Learning.

The first aspect that is covered by CAFLA is collaboration through mobile devices. Our framework allows deploying activities in which mobility of people and devices is possible. Moreover, participants in any activity can collaborate between them. This way, educators will be able to design activities that take into consideration different ways of collaboration, as peer reviewing, quizzes with multiple participants or treasure hunt activities.

In order to make activities more flexible and dynamic, CAFLA allows ubiquitous computing: it means that any participant can work anyplace with others. It is possible to create a collaboration network between participants in a transparent way if they are close enough.

Moreover, contextual information is also present in the framework. Any object in the environment can be tagged by educators with any data they. This feature permits to design learning activities in museums or other environments in which interest objects or spaces can be studied. Contextual information is easy to be received through objects, for example through RFID devices. Moreover, CAFLA includes real time location capabilities. Location is a powerful tool that can be used to cover multiple learning activities and to gather important information that will help to improve learning. For example, art pieces into a museum can provide their information to students if they are in front of them, learning routes can be defined or relevant information about students' collaboration depending on their position can be collected.

The different characteristics and purposes of the activities that CAFLA is able to cover require the implementation of a wireless communication protocol that supports all of them. As we analyze below, ZigBee protocol is the most appropriate to be implemented in the framework.

3.1 Why ZigBee Standard Supports CAFLA?

Due to its nature, ZigBee communication protocol enables to easily deploy communication networks. Forming ad-hoc networks becomes simple and ZigBee devices can change from a network to another in an easy way. Locating ZigBee devices is also possible, as with Wi-Fi technology, but the calibration and infrastructure deploy takes more time with Wi-Fi and even precision is worse. Moreover, energy consumption is lower because ZigBee protocol has been designed for this proposal [21].

Any ZigBee device can be configured to create a network that will be accessible by others to join it quickly and easily anytime and anywhere. This feature allows the creation of ad-hoc networks in a simple way to facilitate, for example, carrying out a collaborative face-to-face activity between two students, collecting data from sensors that provide contextual information or receiving data through devices depending on the placement of students in an environment designed and set in advance. Moreover, we are able to receive the relative position of each node in the network regarding to the other nodes, forming, for example, a map of proximity to each other without requiring a prior infrastructure.

Furthermore, ZigBee networks allow choosing different network topologies: a star topology has an only node that centralizes all the transmissions in the network. In a tree topology there is an element that starts the network and other devices that join it and, at the same time, give other devices access to the network. Finally, a mesh topology allows establishing communication links among all nodes through different routes, being able to get a full connection throughout the network. A mesh topology, together with the large number of devices that can join a network, allows achieving a high scalability, so that peer-to-peer links or networks with large numbers of devices can be formed, covering multiple ways of collaboration among the participants of the activity.

The reliability of ZigBee networks is another feature that benefits learning. The protocol offers fluctuating link capacity that allows sending information between two parties through different paths. Thus, an eventual drop of a link does not block the activity because the data is automatically sent by one of the alternative paths that the network provides. Similarly, any device that falls from a network failure will recover and join the network automatically. This feature allows more dynamic activities as they will be less affected by communication failures. The ease of developing wireless sensor networking using ZigBee is another characteristic that is useful for Context-aware Learning activities. Wireless sensor networks allow gathering contextual information easily through a basic infrastructure that requires a quick and easy deployment. Just connect sensors to a ZigBee device can integrate them into the learning network. This way, designers can think about multiple scenarios in which information is provided to be used by students, as might be collecting data of pollution (air pollution, light and sound) to share and discuss it in groups. Moreover, this infrastructure can be easily move from a place to another without having to be configured again. This simple example allows illustrating a basic idea: the same network will allow collecting data and sending data between devices in any place will be used.

Moreover, ZigBee enables to deploy an infrastructure of devices that transmit data through it and allow locating any ZigBee device moving throughout it. In other words, this technology allows deploying real-time locating systems where individual students or student groups that are developing an activity can be identified and located. These tracking systems identify who is in a concrete area and the proximity of other students who are working face-to-face, mark routes that students should follow, can log all the interactions that have emerged into the network, can determine the most active students or customize the content that the activity must provide depending on the student's location.

From the point of view of energy consumption, the ZigBee communication protocol has been designed to maximize energy saving [21]. This aspect allows developing long-term activities with mobile devices or work in environments where it is not possible to load their batteries (e.g., a park).

Finally, we cannot ignore the potential that these kinds of networks provide for interaction analysis. The network infrastructure deployed with this technology can record the movement of participants, what devices have formed a collaborative network to identify the distance between them, or who has been in an area of interest in a museum, for example. All this information helps to improve the learning process and identify students' different roles within their group, as well as detect and correct deficiencies in the learning design carried out by educators.

4 Conclusions and Future Work

The constant technological advances in today's society have not gone without notice at any social level. The emergence of new technologies, mobile devices and communication protocols and, moreover, its adoption by the majority of the population have allowed that the collection of contextual data, the participation in social networks or the sharing of people's position at a particular time has become as intrinsic to people's daily life as television. In this regard, the field of education has not been immune to these developments and many proposals have been made to include the technology and the its potential in the educational process.

Nowadays, three trends that use mobile devices in this area are the most representative: Mobile Computer Supported Collaborative Learning, Context-aware Learning and Location-based Learning. The technologies that are involved in those kinds of learning usually cover only one of these trends, supporting communication, location or data collection.

However, when designers want to combine different trends or use the characteristics of all of them, two or more wireless technologies should be adopted, in order to support each of the features, getting complicated and non intuitive environments. None of them provides all the requirements a scenario that tries to cover all those kind of learning needs. Therefore, this paper presents CAFLA, an innovate framework that permits developing learning activities that cover the kinds of learning presented before. A brief description of its characteristics is provided and the paper also analyzes the use of ZigBee communication protocol to act as a technological base for the framework presented.

Future work includes a deeper description of CAFLA. On the other hand, the communication protocol working mode and how collaborative networks are created and destroyed must be designed and implemented. In its design, important aspects such as the formation of networks, the identification of users, the location of the site, the capability to integrate with sensor networks and the improvement of the network availability will be taken into account.

Acknowledgments. This project has been supported by the Spanish Ministry of Science and Innovation (Subprograma Torres Quevedo).

References

1. Aarts, E., de Ruyter, B.: New research perspectives on Ambient Intelligence and Smart Environments. Journal of Ambient Intelligence and Smart Environments 1, 5–14 (2009)
2. Roschelle, J.: Unlocking the learning value of wireless mobile devices. Journal of Computer Assisted Learning 19, 260–272 (2003)
3. Sharples, M., Taylor, J., Vavoula, G.: A Theory of Learning for the Mobile Age. In: Bachmair, B. (ed.) Medienbildung in neuen Kulturräumen, pp. 87–99. VS Verlag für Sozialwissenschaften, Wiesbaden (2010)

4. Glahn, C., Börner, D., Specht, M.: Mobile informal learning. In: Brown, E. (ed.) Education in the Wild: Contextual and Location-Based Mobile Learning in Action. A Report from the STELLAR Alpine Rendez-Vous Workshop Series, Nottingham, pp. 28–31 (2010)
5. Koschmann, T.: CSCL: theory and practice of an emerging paradigm. Lawrence Erlbaum, Mahwah (1996)
6. Cortez, C., Nussbaum, M., Santaelices, R., et al.: Teaching science with mobile computer supported collaborative learning (MCSCL). In: Seconds IEEE International Workshop on Wireless and Mobile Technologies in Education (WMTE 2004), pp. 67–74. IEEE Computer Society Press (2004)
7. Trifonova, A., Ronchetti, M.: Hoarding content for mobile learning. International Journal of Mobile Communications 4, 459–476 (2006)
8. Zurita, G., Baloian, N., Baytelman, F.: Using mobile devices to foster social interactions in the classroom, pp. 1041–1046. IEEE (2008)
9. García, Ó., Tapia, D.I., Alonso, R.S., et al.: Ambient intelligence and collaborative e-learning: a new definition model. Journal of Ambient Intelligence and Humanized Computing, 1–9 (2011)
10. Vasiliou, A., Economides, A.A.: Mobile collaborative learning using multicast MANETs. International Journal of Mobile Communications 5, 423 (2007)
11. Neyem, A., Ochoa, S.F., Pino, J.A., Guerrero, L.A.: Sharing Information Resources in Mobile Ad-Hoc Networks. In: Fukś, H., Lukosch, S., Salgado, A.C. (eds.) CRIWG 2005. LNCS, vol. 3706, pp. 351–358. Springer, Heidelberg (2005)
12. Meguerdichian, S., Koushanfar, F., Qu, G., Potkonjak, M.: Exposure in wireless Ad-Hoc sensor networks, pp. 139–150. ACM Press (2001)
13. Laine, T.H., Joy, M.S.: Survey on Context-Aware Pervasive Learning Environments. International Journal of Interactive Mobile Technologies 3, 70–76 (2009)
14. Yang, S.J.H.: Context Aware Ubiquitous Learning Environments for Peer-to-Peer Collaborative Learning. Educational Technology & Society 9, 188–201 (2006)
15. Ogata, H., Yano, Y.: Context-aware support for computer-supported ubiquitous learning. In: 2nd IEEE International Workshop on Wireless and Mobile Technologies in Education (WMTE 2004), pp. 27–34 (2004)
16. Abowd, G.D., Atkenson, C.G., Hong, J., et al.: Cyberguide: A mobile context: aware tour guide. Wireless Networks 3, 421–433 (2006)
17. Curran, K., Norrby, S.: RFID-Enabled Location Determination Within Indoor Environments. International Journal of Ambient Computing and Intelligence 1, 63–86 (2009)
18. Takacs, G., Chandrasekhar, V., Gelfand, N., et al.: Outdoors augmented reality on mobile phone using loxel-based visual feature organization. In: Proceeding of the 1st ACM International Conference on Multimedia Information Retrieval, MIR 2008, pp. 427–435 (2008)
19. De Nardis, L., Di Benedetto, M.-G.: Overview of the IEEE 802.15.4/4a standards for low data rate Wireless Personal Data Networks, pp. 285–289. IEEE (2007)
20. Alonso, R.S., Tapia, D.I., Corchado, J.M.: SYLPH: A Platform for Integrating Heterogeneous Wireless Sensor Networks. International Journal of Ambient Computing and Intelligence 3, 1–15 (2011)
21. ZigBee Alliance, ZigBee 2007 Specification Document (2007),
   http://www.zigbee.org/Standards/Downloads.aspx.

Menu Navigation in Mobile Devices Using the Accelerometer

Alejandro Sánchez, Gabriel Villarrubia, Amparo Jiménez, Amparo Casado,
Carolina Zato, Sara Rodríguez, Ignasi Barri, Edgar Rubión, Eva Vázquez,
Carlos Rebate, José A. Cabo, Joaquín Seco, Jesús Sanz, Javier Bajo,
and Juan Manuel Corchado

Abstract. In this article an application for mobile devices is presented. This
application uses the accelerometer integrated into the own device to detect certain
user movements and use them to navigate through the menus. The application is
destined to those users with visual incapacity who need an alternative mechanism
for the selection of the different options in the menus.

Keywords: accelerometer, menu navigation, mobile devices, disabled people.

1 Introduction

During the recent years there is a growing need for adapting the educational
environments to the new requirements of the information technologies. One of the

Gabriel Villarrubia · Carolina Zato · Sara Rodríguez · Juan Manuel Corchado
Departamento Informática y Automática, Universidad de Salamanca, Salamanca, Spain
e-mail: {gvg,carol_zato,srg,corchado}@usal.es

Alejandro Sánchez · Amparo Jiménez · Amparo Casado · Javier Bajo
Universidad Pontificia de Salamanca, Salamanca, Spain
e-mail: {asanchezyu,ajimenezvi,acasadome,jbajope}@usal.es

Ignasi Barri · Edgar Rubión · Eva Vázquez · Carlos Rebate
INDRA, Spain
e-mail: {ibarriv,erubion,evazquezdeprada,crebate}@indra.es

José A. Cabo
Wellness Telecom, Spain
e-mail: jacabo@wtelecom.es

Joaquín Seco · Jesús Sanz
CSA, Spain
e-mail: {joaquin.seco,jesus.sanz}@csa.es

P. Vittorini et al. (Eds.): International Workshop on Evidence-Based TEL, AISC 152, pp. 133–140.
springerlink.com   © Springer-Verlag Berlin Heidelberg 2012

segments of the population which will benefit with the advent of systems based on Ambient Intelligence will be the elderly and people with disabilities [6], contributing to improve their quality of life 7. Ambient Intelligence evolves from the ubiquitous computing 3, and constitutes the most promising technological approach to meet the challenge of developing strategies in dependency environments [8].

This work presents an innovative method, based on the Ambient Intelligence (AmI) paradigm [4, 7], for formal teaching of languages oriented to disabled people, specifically, with visual disabilities. In this way, it is proposed a method to facilitate the integration of disabled people in the information society. The special characteristics of disabled people require new solutions, since the learning methods used in early ages are most of the times not appropriated for this sector of the society. It is necessary to investigate in new techniques and methods to satisfy the learning needs of the elderly people.

This paper focuses in the combination of the new information technologies along with the traditional teaching. In this way it will be possible to combine the advantages of traditional teaching and the advantages of the mobile devices. It will be necessary to upgrade the systems of evaluation/accreditation to assess the knowledge or skills acquired during the learning process. To achieve this objective, we propose the use of mobile devices, intelligent systems and wireless communications.

The rest of the paper is structured as follows: Next section introduces the problem that motivates most of this research. Section 3 presents the proposed application for mobile devices used by visual disabled people. Finally, section 4 explains some conclusions obtained.

2 Background

At the moment, numerous technological advances exist to facilitate the daily workings to the disabled or elderly users, adapting the technology to their needs. For it, in recent years, the use of smartphones is growing due to the wide range of possibilities that these offer. This type of mobile telephones includes a series of complements like a-gps, radio, digital compass or the accelerometer.

There is an ever growing need to supply constant care and support to the disabled and elderly and the drive to find more effective ways to provide such care has become a major challenge for the scientific community [3]. During the last three decades the number of Europeans over 60 years old has risen by about 50%. Today they represent more than 25% of the population and it is estimated that in 20 years this percentage will rise to one third of the population, meaning 100 millions of citizens [3]. In the USA, people over 65 years old are the fastest growing segment of the population [1] and it is expected that in 2020 they will represent about 1 of 6 citizens totaling 69 million by 2030. Furthermore, over 20% of people over 85 years old have a limited capacity for independent living, requiring continuous monitoring and daily care [2]. Some estimations of the World Health Organization show that in 2025 there will be more than 1000 million people aged over 60 in the world, so if this trend continues, by 2050 will be double, with about the 80% concentrated in developed countries [9].

Education is the cornerstone of any society and it is the base of most of the values and characteristics of that society. The new knowledge society offers significant opportunities for AmI applications, especially in the fields of education and learning [8]. The new communication technologies propose a new paradigm focused on integrating learning techniques based on active learning (learning by doing things, exchange of information with other users and the sharing of resources), with techniques based on passive learning (learning by seeing and hearing, Montessori, etc.) [5]. While the traditional paradigm, based on a model focused on face to face education, sets as fundamental teaching method the role of the teachers and their knowledge, the paradigm based on a learning model highlights the role of the students. In this second paradigm the students play an active role, and build, according to a personalized action plan, their own knowledge. Moreover, they can establish their own work rhythm and style. The active methodology proposes learning with all senses (sight, hearing, touch, smell and taste), learn through all possible methods (school, networking, etc.), and have access to knowledge without space or time restrictions (anywhere and at any time).

There are different studies that have used the Ambient Intelligence to facilitate learning. In [3], Bomsdorf shows the need to adapt intelligent environments to changes depending on the educational context and the characteristics of users. Morken *et al.* [9] analyze the characteristics of intelligent environments for learning. They focus on the role of mobility in educational environments and the role that acquire the mobile devices. Naismith *et al.* [10] conducted a detailed study describing the role of mobile devices in education, analyzing the characteristics of the devices and their capacity for learning in educational environments. All these approaches are focused on the role of learning in Ambient Intelligence environments, but none of them is oriented on learning for dependents or elderly people. The following section presents a multiagent architecture that facilitates learning methodology using an active through mobile devices.

2.1 Operation of an Accelerometer

An accelerometer is an electromechanical device that measures the forces of the acceleration [24]. The accelerometer provides new possibilities of interaction with the users, allowing to place orders by movements. These devices belong to category MEMS (Micro Electro-Mechanical Systems), which are a type of electromechanical devices, constructed normally with remodeled polycrystalline silicon.

The forces that measure can be:

- Static forces: Gravity towards the Earth center
- Dynamic forces: Movement or the vibration of the devices

The operation of the accelerometer is relatively simple. First of all, it measures the static acceleration of the gravity, obtaining therefore the angle of the device with respect to the Earth. Next, it measures with the sensors of dynamic acceleration. The movement of the accelerometer can be analyzed in the three dimensions.

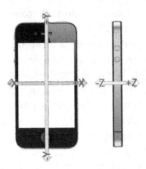

Fig. 1 Dimension of the operation in accelerometers.

2.2 *Devices with Accelerometer*

There are a great amount of devices that own an accelerometer within them. It surprises now that the first accelerometers were not those included in the devices of mobile telephony, they were used previously. One of the first uses of this apparatus was in hard disks. Its function consisted on measuring the vibrations and inclination of the device when changing of internal disc and thus to try that there were not problems of internal collisions.

Another use of the accelerometer was in the last generation of games consoles. Nintendo released the novel Wii. This game console broke the concepts previously established, since the controls are not connected, they communicated via infrared, having in addition, an accelerometer to catch the movements of the user. In this section, the mobile telephony can be also included, since the majority of devices in this scope, only uses it as part of entertainment applications.

Finally, a research group of the Pontifical University of Salamanca carried out a project, called AESCOLAPIUS, for the construction of a wheelchair able to move using the different functionalities that smartphones offer. Within the scope our work the most interesting functionality is the offered by the accelerometer, since the chair reacts to the movements made with the device [11].

3 Developed Application

The application is based on a navigation menu using the accelerometer available in many mobile devices, especially in the iPhone platform. This application may solve the problems that visually impaired people have while using a menu with different options. Cutting-edge devices have touch screens and it's impossible for them to identify where they are pressing.

A movement recognition algorithm has been created in order to collect all data necessary for the successful operation of the system.

The designed algorithm is based on two distinct aspects: the current position of the device and its previous position. The time gap used between both measurements is one second, being this gap fully configurable. After multiple tests, it was concluded that a shorter gap affects the stability of the application, since the values of the accelerometer are constantly changing. As seen above, the device is able to move in three different axes, being the variation in each of them a value between 1 and -1. Next, movements, which are taken into account in the algorithm, will be defined.

Central Position
This will be fulfilled when previous X has values between 0,2 and -0,2 and present X has values between 0,2 and -0.2

Movements to the Right
A turn limit has been settled down to detect when a movement to the right takes place, this movement is necessary to navigate through one option of the menu. This limit is the previous position plus a turn rank, which has settled down that is needed to turn at least 45° towards the right. This movement needs that the device comes from the position centered in x-axis.

This will be fulfilled when present X is major that previous X the 0,5 and device is centered more in X.

Cancelation of an Option
The elimination of an option to be able to choose another one of the menu, is carried out by a left turn of the device, starting the movement from the vertical position. If an option is eliminated, doing a movement to the right the following option of the menu is selected.

This will be fulfilled when previous X is major that present X 0,5 (value limits) and we are trims in X more.

Access control of an Option
The selection of an option is the fundamental pillar of the algorithm, since the values taken with the previous functions are used. The access to an option of the menu is carried out through movements on z-axis (movements of back forwards or vice versa). In order to enter an option, this need to be previous selected by movements to the right and left, and then, make a movement with the device to backwards.

The access to an option will be fulfilled when we are centered in X and Z axis and we have chosen an option previously (without cancelled it) and that previous value Z is major that present Z plus 0,5 (value limits).

Return to the Previous Menu
The exit to a previous menu will allow the user to leave a screen when it does not need it, and to return to a previous menu.

The exit will be fulfilled when we are centered in the X and Z axis and that present value Z is major that previous Z plus 0,5 (value limits)

```
IF (0.2 >= X_previous <= -0.2) and (0.2 >= X_current <= -0.2)
     THEN centeredInX = true ---> Central position
IF (0.2 >= Z_previous <= -0.2) and (0.2 >= Z_current <= -0.2)
     THEN centeredInZ = true ---> Central position
IF (X_current >= (X_previous +0.5) and centeredInX)
     THEN selectedChoice = true ---> Right movement
IF (X_previous >= (X_current +0.5) and centeredInX)
     THEN selectedChoice = false ---> Option cancelation
 IF (centeredInX and centeredInZ and selectedChoice and (Z_previous >=
(Z_current +0.5)))
          THEN goToMenuOption ---> Option acceptation
 IF (centeredInX and centeredInZ and (Z_current >= (Z_previous +0.5)))
     THEN goBack()---> Go back
```

Fig. 2 Algorithm for controlling the movements

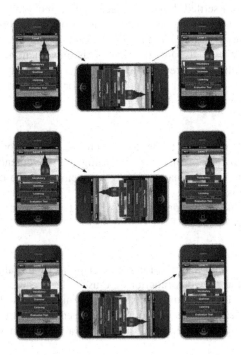

Fig. 3 Operation example

4 Conclusions

The project focus on the learning of languages through mobile telephone, in our case iPhone, and more in particular, oriented to people with visual disability.

This paper has presented a new method for teaching languages to disabled people, using techniques of Ambient Intelligence and mobile devices. The new algorithm, based on the use of the accelerometer, presented in this paper provides an active learning through simple and intuitive interfaces, installed on mobile devices and movement recognition. This requires the integration of intelligent algorithms with innovative strategies of teaching languages and mobile devices. In this way we have obtained:

- An active method of teaching and learning of languages for people with visual disabilities.
- An interaction system based on the AmI paradigm for language education.
- The obtained language learning system was adapted to be executed on mobile devices, facilitating the adaptation to the needs of the dependent and elderly people.

The system has been developed under the framework of the AZTECA project and has been tested by the visually disabled users involved in this project. The users have remarked the potential possibilities of the interaction method. However, it is still necessary to perform a more general evaluation with final users.

Acknowledgements. This project has been supported by the Spanish CDTI. Proyecto de Cooperación Interempresas. IDI-20110343, IDI-20110344, IDI-20110345, and the MICINN TIN 2009-13839-C03-03 project. Project supported by FEDER funds.

References

1. Anastasopoulos, M., Niebuhr, D., Bartelt, C., Koch, J., Rausch, A.: Towards a Reference Middleware Architecture for Ambient Intelligence Systems. In: ACM Conference on Object-Oriented Programming, Systems, Languages, and Applications (2005)
2. Angulo, C., Tellez, R.: Distributed Intelligence for smart home appliances. In: Tendencias de la Minería de Datos en España, Red Española de Minería de Datos, Barcelona, España (2004)
3. Bajo, J., Corchado, J.M., de Paz, Y., de Paz, J.F., Rodríguez, S., Martín, A., Abraham, A.: SHOMAS: Intelligent Guidance and Suggestions in Shopping Centres. Applied Soft Computing 9(2), 851–862 (2009)
4. Bajo, J., de Paz, J.F., de Paz, Y., Corchado, J.M.: Integrating Case-based Planning and RPTW Neural Networks to Construct an Intelligent Environment for Health Care. Expert Systems with Applications 36(6), Part 2, 5844–5858 (2009)
5. Brown, T.H.: Beyond constructivism: Exploring future learning paradigms. In: Education Today, vol. (2). Aries Publishing Company, Thames (2005)

6. Carretero, N., Bermejo, A.B.: Inteligencia Ambiental. CEDITEC: Centro de Difusión de Tecnologías. Universidad Politécnica de Madrid, España (2005)
7. Echt, K.V.: Designing web-based health information for older adults: visual considerations and design directives. In: Morrell, R.W. (ed.) Older Adults, Health Information and the World Wide Web, pp. 61–87. Lawrence Erlbaum Associates (2002)
8. Friedewald, M., Da Costa, O.: Science and Technology Roadmapping: Ambient Intelligence in Everyday Life (AmI@Life). Working Paper. Institute for Prospective Technology Studies IPTS, Seville (2003)
9. Kurniawan, S.H., King, A., Evans, D.G., Blenkhorn, P.L.: Personalising web page presentation for older people. Interacting with Computers 18, 457–477 (2006)
10. Naismith, L., Lonsdale, P., Vavoula, G., Sharples, M.: Futurelab Literature Review in Mobile Technologies and Learning. Technical Report for Futurelab (2004), http://www.futurelab.org.uk/research/reviews/reviews_11_and12/11_01.htm
11. Berjón, R., Mateos, M., López, A., Muriel, I., Villarrubia, G.: Alternative human-machine interface system for powered wheelchairs. In: Corchado, J.M., Bajo Pérez, J., Hallenborg, K., Golinska, P., Corchuelo, R. (eds.) Trends in Practical Applications of Agents and Multiagent Systems. AISC, vol. 90, pp. 307–315 (2011)
12. Bao, M.H.: Micro Mechanical Tansducers: Pressure Sensors, Accelerometers and Gyroscopes. Handbook of Sensors and Actuators, vol. 8. Elsevier Editions

Author Index